PRACTICAL ASSESSMENT

BIOLOGY

for GCSE

James Lewis

UNWIN HYMAN

Published by
UNWIN HYMAN LIMITED
15–17 Broadwick Street
London W1V 1FP

British Library Cataloguing in Publication Data

Lewis, James
    Practical Assessment: Biology for G.C.S.E.
    1. Biology——Study and teaching
    I. Title
    574′.024372   QH315

ISBN 0-7135-2824-9

Illustrated by RDL Artset Ltd
Cover design by Charles Snape

Typeset by Pentacor Ltd., High Wycombe, Bucks
Printed in Great Britain by Bell & Bain Ltd., Glasgow

Contents

Acknowledgements

The author thanks Professor David Waddington of York University for his help in furthering the publication of this book; Chris Blake and Pat Winter of Unwin Hyman Limited for their help and advice; and Carmel his wife, for her endless patience.

1 Why internal assessment?

The National Criteria for Science provide an agreed framework for GCSE courses. They require that:

"All schemes of assessment must allocate not less than 20% of the total marks to practical skills. At least half of these marks must be awarded on the basis of experimental and observational work in the laboratory or its equivalent."

"Some of the associated skills can be assessed by means of questions in a written paper and such questions could supplement, but not replace, assessment based on work within the laboratory or its equivalent. For internal candidates, the internal assessment of course work moderated by the Examining Groups is likely to be the most appropriate approach and formal practical tests, unsupported by coursework, would not normally be regarded as acceptable."

Formal practical examinations have been rejected as a means of assessing practical skills because:

"(They) tend to narrow the range of experimental work carried out during a course. They can encourage repetitive, even trivial, exercises. . . ."

The requirement that practical skills be assessed internally is the aspect of GCSE Biology which has caused teachers most anxiety. Concern over lack of experience in the internal assessment of skills is compounded by worries over large classes, lack of equipment, lack of technical support and disruptive pupil behaviour. This guide and sourcebook has been written with these very real problems uppermost in mind.

Teachers who have already operated schemes of internal assessment, either in courses at 0/CSE level or in A–level courses as an alternative to a practical examination, have found that there are benefits to be gained from the internal assessment of practical skills. At its best internal assessment can:

– increase teachers' understanding of the purpose of practical work and improve its use.

– provide teachers with feedback on their students work.

– allow a wide range of tasks to be used as the basis for the assessment without distorting the coursework unduly.

– increase pupils' motivation and commitment in practical work.

– allow assessment to take place in the normal course of practical work, without the need to set up special test situations.

– test the possession of manipulative, observational and measurement skills by examining the process and not just the outcome of experimental work.

– allow differentiated assessment.

– minimise the stress that some candidates suffer in the formal setting of practical examinations.

– avoid the problems inherent in having to perform at the highest possible level on the day in formal examinations.

This guide and sourcebook provides suggestions for drawing up a scheme of internal assessment of practical skills, and suggestions about the management of assessment of practical skills. Chapter 5 includes 20 exemplar exercises which illustrate the principles discussed in the earlier chapters.

2 What needs to be assessed?

Introduction

The National Criteria for Biology include the following assessment objectives which relate to experimental work and which examining groups could therefore be expected to develop schemes to assess.

Candidates should be able to:

3.2.1 make and record accurate observations;
3.2.2 plan and conduct simple experiments to test given hypotheses;
3.2.3 formulate hypotheses and design and conduct simple experiments to test them;
3.2.4 make constructive criticisms of the design of experiments;
3.2.5 analyse, interpret and draw inferences from a variety of forms of information including the results of experiments.

These criteria are open to a range of interpretations because some of them (eg 3.2.1) encompass more than one practical skill and some (eg 3.2.2) are vague in their requirements. This may explain why it is that the five examining groups have devised such a variety of different schemes. The schemes of assessment of the five groups are compared with each other and with the National Criteria objectives in Table 2.1.

Comparison of these schemes is interesting. Firstly all of the examining groups have agreed that some of the National Criteria objectives involve the grouping together of several objectives which they clearly feel should be assessed separately. Objective 3.2.1: "Make and record observations" involves observation, measurement and recording and this has been recognised by all of the groups, although to differing degrees. All (except the NEA) group two of these objectives together; however, which two are grouped varies. The

Table 2.1   Summary of Experimental Skills Assessment Schemes

National Criteria for Biology	NEA	SEG	MEG	LEAG	WJEC
3.2.1. Make and record accurate observations	1 Measurement (1–5) 2 Observation (6–10) 4 Recording (15–19)	3 Make and convey accurate observations 4 Record results in an orderly manner.	3 Observing and measuring 4 Recording and communicating	A Make and record accurate observations B (ii) Record of results	1 Observational and Recording skills. 2 Measurement skills.
3.2.2. Plan and conduct simple experiments to test hypotheses	3 Handling apparatus and material (11–14)	1 Follow instructions 2 Handle apparatus and materials 7 Safe working procedures	1 Following instructions 2 Handling apparatus and materials	B (i) Performing experiments	3 Procedural skills 2 Manipulative skills.
3.2.3. Formulate hypotheses and design & conduct simple experiments to test them	6.2 Design (29–32)	5 Formulate a hypothesis 6 Design an experiment to test a hypothesis.	6 Experimental design/problem solving	C Designing and evaluating an experiment (i) Plan of investigation	5 Formulate a hypothesis and design and conduct an experiment to test it.
3.2.4. Make constructive criticisms of the design of experiments	6.1 Design (27–28)	Assessed by written examination	6 (e) Evaluate methods and suggest improvements	C Designing & evaluating (ii) Evaluation of experiments.	Assessed by written examination.
3.2.5. Analyse, interpret and draw inferences from results.	5 Data and its interpretation (20–26)	Assessed by written examination	5 Interpreting data	B (ii) Interpretating of results	Assessed by written examination.
Weighing	30%	20%	20%	20%	20%
No. of Assessments	As many as are needed to demonstrate competence	1–4 : 3 times 5, 6 : 1	1–5 : at least 2 6 : 1	A, B : 2 ; C : 1	1–4 : at least 2 5 : at least 1
Scale used	0/1	H, J, K, L	0/9	0/5	1–4 : 0–3 5 : 0–3 (doubled)
Moderation procedure	By inspection	Moderation by inspection of marks and lists of exercises.	Internal moderation; External moderation of sample of work	External inspection of all record sheets & sample of work	External moderation of sample of work
Maximum mark	32	N/A	99	40	30
Guidance on activities to be used for assessment	May use special exercises	**Must** be during normal practical work	1–3 based on pupil's practical work; 4–6 on written work	Some suited to practical test, others as normal practical work	**Avoid** formal practical test; part of normal practical work
Other guidance	Assessment only carried out once competence is known	Normally made within 1 year of examination	Marks must represent candidate **best** performance	Assessed during second half of course	Not more than 2 skills assessed on one occasion

NEA, SEG and MEG also separate objective 3.2.3: "Formulate hypotheses and design and conduct simple experiments to test them" into more than one assessment objective. Grouping objectives such as measurement and observation, or observation, and recording recognises that it is very difficult to isolate one of them and assess it without the pupil having competence in the other skill. This approach also has the advantage that assessment of objectives, within the course of normal practical work, becomes a more credible proposition if there are less objectives to assess because some of them have been grouped together.

In many other respects the schemes drawn up by the examining groups are similar: the number of assessments, the method of moderation, the weighting of the internal assessment.

From assessment objectives to performance criteria

Having chosen which syllabus to use, a biology teacher or department is then faced with the task of drawing up a practicable framework of assessment which can be incorporated into normal coursework. All except the NEA scheme list the objectives they wish assessed but give little detail of what should actually be assessed, or definition of what competence at different levels of attainment should be. For example, when attempting to assess the objective "Handling materials and apparatus", what apparatus should a low, medium or high attaining pupil be able to handle? What constitutes "success" in handling the apparatus? To put into practice any of these four schemes, it is first necessary to draw up a set of performance criteria against which pupils can be measured. These four schemes have done this to different extents but the performance criteria are often vague. Definitions of both the range of skills over which assessment will take place and the performance criteria for each skill within an assessment objective, must therefore be drawn up before the schemes can be put into practice. Every scheme is different and the skills to be included in the performance criteria for each assessment objective depend upon the equipment and conditions prevailing in individual departments. Nevertheless, examples of the range of skills which could be included in the performance criteria for each objective are given in subsequent sections where assessment objectives are considered individually.

The NEA scheme is a special case because they have already defined very precisely the skills which they wish to have assessed. They have also defined with reasonable clarity what constitutes success in each of these skills, thus it is not necessary for teachers using this scheme to draw up performance criteria for the skills to be assessed. The teacher has to make the decision as to whether or not success has been demonstrated and award a "tick" or "cross" (or 0/1) for that pupil, rather than deciding which level of performance on a scale the pupil has reached, as is the case with the other four schemes.

In the remainder of this chapter consideration is given to drawing up performance criteria for each objective to be assessed. For each objective a set of performance criteria for each of three levels of attainment, has been suggested to illustrate how performance

criteria may be framed. The MEG and WJEC schemes use this system but the SEG and LEAG schemes make use of 4– and 5–point scales respectively. Thus these schemes would require performance criteria to be drawn up at each of these levels of attainment. To provide a fuller definition of each of the objectives, the discussion draws on the assessment schemes of all five examining groups and the 1985 Report of the Working Party for Biology established by the Secondary Examinations Council, to propose Draft Grade Criteria at GCSE.

Since the four schemes to be considered differ in the objectives to be assessed, a common set of objectives has been drawn up for this purpose. The objectives from the National Criteria for Biology listed above are unsuitable because they group together objectives which some schemes have separated, and because some of the objectives are quite vague. Thus a list of seven objectives common to all schemes has been used. This list is as follows and will be used in the remainder of this book:
Objective 1: Following instructions
Objective 2: Handling apparatus and materials
Objective 3: Observation
Objective 4: Measurement
Objective 5: Recording
Objective 6: Interpretation of results
Objective 7: Designing an experiment to test a hypothesis.

Objective 1: The ability to follow instructions

Some of the syllabi make it clear that the instructions can be given in written, diagrammatic or oral form. The problems here are to decide what range of manipulations might pupils be expected to carry out in the course of following instructions and on what basis discrimination can be made between different levels of attainment. Ability to follow instructions can be confused with ability to manipulate apparatus; it is likely that unless adequate care is taken to ensure that this does not happen pupils will attain lower than their true level because the apparatus is difficult to manipulate. Procedures used in assessing this skill must therefore be simple enough for the lowest attainer to be capable of carrying out.

Discrimination may be achieved by increasing the complexity of the sequence of operations to be carried out. The emphasis in all the schemes, however, is on discrimination on the basis of degree of assistance required in following the instructions. But how can this be quantified? It would be possible and desirable to draw up a set of performance criteria which expresses competence at each level in terms of the complexity of the sequence of instructions carried out without assistance.

At the lowest level, the assessments might test pupils' ability to follow simple instructions one at a time, for example add 3 drops of iodine solution to the starch solution in the test tube. At a higher level, the pupils should be able to follow a sequence of instructions. At the highest level, they might be expected to follow a sequence of branched instructions, for example add Biuret solutions to the solution of food in the test-tube. If a violet colour appears test another food in the same way. If there is no violet colour, add Benedict's solution to the same food solution and heat gently, looking for an orange/red precipitate.)

Table 2.2 shows how the points listed above could be used as the basis for performance criteria. The terms "high", "medium" and "low" would clearly be replaced by the grades used in which ever scheme is being used.

Table 2.2

Objective 1: The ability to follow instructions	
High	Able to follow a sequence of instructions without help including branched instructions.
Medium	Able to follow a sequence of instructions without help.
Low	Able to follow a single written, diagrammatic verbal instruction.

Objective 2: The ability to handle apparatus and materials

A problem to be faced in drawing up the performance criteria is to draw up a list of apparatus which can be used for assessment; apparatus that a department expects GCSE pupils to be able to manipulate correctly. As with Objective 1, the contents of this list will depend upon the equipment in use in the normal course of GCSE Biology work in a particular department. An example of such a list is given below.

Apparatus to be used for assessment of the objective "Handling apparatus and materials".

Candidates should be able to confidently, correctly and safely use each of the following:

- ● Microscope
  Hand lens
  Dissecting instruments
- ● Simple potometer
- ● Simple osmometer
- ○ Test tubes and racks
- ● Simple respirometer
- ○ Stopclock
  Clamp stand, clamp and boss
  Bunsen burner
- ○ Spatula
- ○ Tongs
- ○ Safety glasses/goggles
- ○ Beaker
- ○ Tripod
- ○ Gauze
  Measuring cylinder
  Balance
  Thermometer

● = apparatus requiring a high level of competence to manipulate successfully.

○ = apparatus requiring a low level of competence to manipulate successfully.
(See discussion below)

It is then necessary to define competence in manipulating each piece of apparatus. This is illustrated with the bunsen burner and simple potometer:

Bunsen burner
- ensure rubber tubing is fully pushed onto gas tap
- have lighter ready when lighting
- collar closed when lighting and when not in use
- turn gas tap on after above precautions have been taken
- adjust temperature using gas tap
- safely positioned on bench mat

Simple potometer
- remove leafy shoot from stem and keep cut end under water
- do not submerge lamina
- fill capillary tubing with water
- insert leafy shoot in capillary tubing without allowing air into column of water
- joint sealed with petroleum jelly
- potometer firmly clamped in boss, clamp and stand
- 0.5–1 cm long air bubble drawn in at bottom of potometer

Successful manipulation of some of these pieces of apparatus would be expected from all pupils but a number of items are applicable only to pupils of higher ability. The examples of the bunsen burner and simple potometer above are chosen to illustrate this; the bunsen burner is likely to be a piece of apparatus which requires a lower level of ability to successfully manipulate than a simple potometer. Thus to allow discrimination it is necessary to identify the level of competence required to successfully manipulate each item in the list. In deciding how difficult manipulation of a piece of apparatus is the number of steps required in the successful assembly of the apparatus and/or the ease of manipulation required, could be taken into consideration. The bunsen burner has almost as many steps in its use as the assembly of the potometer but because the manipulation involved in the use of the bunsen burner is less demanding, a low or medium attaining pupil might be one that could successfully manipulate the bunsen burner but not the potometer. The 'o' sign indicates which items in the list a low achieving pupil might be expected to manipulate. The '●' sign indicates those items which only the high achieving pupils might be expected to manipulate.

All of the syllabi stress the safe use of apparatus in one way or another. It is usually included as part of this assessment objective, although the SEG devote an assessment objective entirely to safety. However this objective is given an impression mark for safe working throughout the whole course by the SEG. It could therefore be incorporated as a skill within the objective "handling apparatus and materials" as other syllabi have done and at the end of the course a grade given for the level of success in the safety objective based on the marks achieved for safe use of apparatus when assessed as a skill within the objective of "handling apparatus and materials". This has the advantage of raising the level of objectivity of the assessment of this objective.

Table 2.3 illustrates the way in which performance criteria for the objective "handling apparatus and materials" could be more tightly defined by listing the apparatus which might be expected to be used successfully, defining competence in each item, and producing an order of difficulty.

Table 2.3

Objective 2:	The ability to manipulate apparatus and materials
High	Able to use all of the apparatus in the list correctly and safely.
Medium	Able to use all of the apparatus in the list except those items for which it has been agreed an advanced level of manipulative skill is required.
Low	Able to use simple items of apparatus in the list, eg forceps, thermometer.

Table 2.4

Objective 3:	The ability to observe
High	Able to observe all relevant gross and fine features and differences in a given situation without cueing.
Medium	Able to observe a range of relevant gross features and fine details in a given situation in response to general cues.
Low	Able to make an observation in response to a specific cue.

Objective 3: The ability to observe

Here again, it is necessary to decide what skills make up this objective. These are listed below:

- The ability to match a specimen to the correct one from a range of examples. (eg matching a particular beetle to one in a set of examples of closely related beetles)
- The ability to observe gross features
- The ability to observe fine detail
- The ability to observe differences in gross features
- The ability to observe differences in fine detail
- The ability to observe changes

The Draft Grade Criteria for Biology suggest that discrimination is on the basis of the amount of cueing pupils are given. A low achieving pupil would thus be able to make an observation if given a clear indication of what to look for. For example given a sugar solution boiled with Benedict's reagent, observe the colour of the resulting precipitate. A high attaining pupil is given a sugar solution with Benedict's reagent and told to report the result as the solution is boiled.

In addition to a consideration of cueing it is possible to discriminate using the difficulty of the skills: observation of fine details requires a higher level of competence in the objective than does the observation of gross features. (Fine details are defined as details of a gross feature eg when observing a cow skull, a gross feature is that it has teeth, a fine detail would be the arrangement, colour, size and number of teeth.)

A further point for consideration is that it is possible to discriminate depending on the quantity of observations made. A low attaining pupil would be able to observe a single colour change in a solution, or a single gross feature of a specimen. However if there was a sequence of colour changes or several gross features, a higher level of competence would be needed to observe these. Under these circumstances a low attaining pupil would only observe a small proportion of the features or changes to be observed.

Table 2.4 shows how performance criteria for this objective could be drawn up based upon the relative difficulty of the skills listed above, the completeness of the observations and on the amount of cueing provided for a pupil rather than merely the number of each type of observation.

Objective 4: The ability to measure

As was the case in the objective of handling apparatus, a list of measuring instruments which pupils at GCSE level should be able to use must first be drawn up. Some guidance in doing this is given by some syllabi since they specify which quantities pupils should be able to measure. They might be the measurement of mass, length, volume, time and temperature, as suggested by the LEAG scheme. Thus the instruments to be used would be these:

- thermometer
- metre rule
- measuring cylinder
- balance (which ever sort is in normal laboratory use)
- stopclock (which ever sort is in normal laboratory use)

The required accuracy of measurement would be the nearest division on a scale. At the lowest level of achievement pupils would be expected to read a scale already in position to the nearest division on a whole number scale, (eg read the temperature of a water bath using a thermometer which is in place in the beaker). At the highest level of achievement multiple or fractional scales would be used and pupils would again be expected to read to the nearest division. (Eg measure out 5.6 cm^3 of water using a 0–10 cm^3 measuring cylinder.)

Table 2.5 shows how performance criteria for this objective might be drawn up.

Table 2.5

Objective 4:	The ability to measure
High	Can measure with no error using all of the instruments in the list when provided with the instrument and the quantity to be measured. Can do so using instruments with fractional or multiple scales.
Medium	Can measure with no error using all of the instruments in the list with whole number scales when provided with the instrument and the quantity to be measured.
Low	Can measure with no error using all of the instruments in the list with whole number scales when the instrument is placed in position.

Objective 5: The ability to record results

Where any guidance is given for this objective examining groups agree that pupils should be assessed on their ability to record results in tables, point graphs, bar graphs and drawings/diagrams. Guidance on how to discriminate can be found in the Draft Grade Criteria for Biology. This is likely on the basis of whether or not pupils can draw up a framework for the recording of results, whether or not they can fill in a framework accurately and comprehensively whether or not they take account of correct units in their recordings and the relative neatness of their recordings.

Table 2.6 shows possible performance criteria for this objective.

Table 2.6

Objective 5:	The ability to record results
High	Able to record results in neat tables with appropriate headings and units. Able to draw neat, accurate, clear and fully labelled diagrams to record observations. Able to plot labelled point and bar graphs choosing suitable axes and scales and where decimal and negative values are involved.
Medium	Able to record results in tables which are labelled sufficiently clearly to show what the results are. Able to draw simple, neat labelled diagrams to record observations. Able to draw and label bar graphs and point graphs where positive whole numbers are involved.
Low	Able to record results in a given format: prepare tables, bar graphs and point graphs involving positive whole numbers. Able to record observations in a simple labelled diagram sufficiently clearly to show what the observations are.

Objective 6: The ability to interpret results

Table 2.1 shows that two boards have chosen to assess this objective entirely on the end of course written examination, indicating the intellectual nature of the objective. This objective includes the following skills:

- the ability to make simple calculations involving data (for instance averages and percentages);
- the ability to extract information from results; interpolation and extrapolation of graphs, as well as reading corresponding values from a table;
- the ability to recognise patterns in data;
- the ability to draw valid conclusions from data;
- the ability to explain a pattern or conclusion in terms of underlying theory.

Some of the skills listed above are more difficult than others (eg extrapolation is a more complex task

than interpolation) thus providing a basis for discrimination. Discrimination can also involve the thoroughness of the conclusion and explanation of results given. At a low level pupils might be expected to be able to draw a single, simple conclusion from a result; for example the orange/red colour indicates that sugar is present in the solution. At a higher level pupils might be expected to recognise a pattern; for example as the temperature increases the speed at which the fungus multiplies increases. At the highest level pupils should be able to give an explanation of why the fungus grows faster as the temperature increases (provided, of course, they have been taught sufficient theory to be able to explain this).

Table 2.7 shows possible performance criteria for this objective.

Table 2.7

Objective 6:	The ability to interpret results
High	Able to identify patterns and draw conclusions from results and explain them in terms of fundamental principles. Able to interpolate and extrapolate from graphs and read corresponding values from tables. Able to make simple calculations with data including percentages and averages.
Medium	Able to identify patterns and draw conclusions from results. Able to interpolate from graphs and read corresponding values from tables. Able to make simple calculations with data including percentages and averages.
Low	Able to identify a simple pattern and draw a conclusion from a single result. Able to read corresponding values from tables. Able to make simple calculations with data including averages.

Objective 7: The ability to design an experiment to test a hypothesis

All of the schemes refer in some way to the need to first formulate a testable hypothesis and then draw up the plan of the investigation which tests it. The plan of the investigation must involve statement of the variables which have to be controlled, how they can be controlled, the procedure which is needed to test the variable under investigation and all of the measurements/observations which must be made in testing the hypothesis. At a low level pupils may be able to suggest a single stage experiment to test a given hypothesis, taking account of some of the variables and suggesting, in outline, the procedure required. For example, when given the hypothesis that acid will diffuse into the centre of a small piece of jelly more quickly than a large piece of jelly, pupils might be able to suggest the equipment required and state that the amount of acid added would have to be the same, that the stopclock must be started at exactly the same moment in each test, and the time taken for the piece of jelly to change from red to yellow recorded in each case. At a higher level pupils could be expected to suggest a way of speeding up the diffusion of acid into the centre of a large block. They would suggest testable hypotheses, such as increasing the concentration of acid, or reducing the size of the cube (the logic of the hypothesis is unimportant as long as it is a hypothesis that can be tested). Pupils would then be able to design an experiment to test their hypothesis taking account of a range of variables, taking steps to control them and noting the measurements required. Table 2.8 shows possible performance criteria for this objective.

Table 2.8

Objective 7:	The ability to design an experiment to test a hypothesis
High	Able to suggest hypotheses which can be tested by experiment. Able to plan a suitable procedure to test a hypothesis taking account of all variables to be controlled and suggesting adequate ways of controlling them. Able to suggest all measurements/observations to be recorded.
Medium	Able to suggest a hypothesis which can be tested by experiment. Able to plan a suitable procedure to test the hypothesis taking account of some variables and suggesting reasonable ways of controlling them. Able to suggest all measurements/observations to be recorded.
Low	Able to plan a suitable simple procedure to test a given hypothesis. Able to suggest some variables which must be controlled. Able to suggest measurements/observations which must be recorded in this simple experiment.

3 Managing the assessment

Introduction

For many teachers the internal assessment of practical work is, first and foremost, a management problem. Problems of equipment shortage, lack of technician support, large class sizes and disruptive pupils, may occur in various combinations in very many schools. Under such circumstances management of practical work itself is difficult enough let alone the assessment of skills. Nevertheless assessment is necessary, and this chapter gives consideration to some strategies which can help to overcome some or all of these difficulties.

What follows is a range of different management strategies. They could be used as they are or modified to the particular circumstances which prevail in the department. They differ in the amount of student practical work they involve and therefore the National Criteria assessment objectives to which they are appropriate. They also differ in the degree of difficulty, preparation and energy required in carrying them out. Examples of all of these strategies are included in the exercises in Chapter 5.

Questions requiring written answers

For many years written examinations have included questions which test pupils' practical skills. Written exercises which are less formal could be used more extensively than examinations. They could be used in the context of a normal course of practical work, or in end of topic tests. Of the seven objectives it could be argued that all but the ability to follow instructions for practical work may be assessed using this approach.

This approach has the immediate appeal of being so easy to manage, both in terms of assessing a large number of pupils at once and in preventing collusion between pupils. In addition it makes no demands on apparatus or technician support, all pupils are presented with the same exercise (thus ensuring fairness) and it is relatively easy to mark. This approach has been adopted by the SEG and WJEC to assess pupils' ability "to make constructive criticisms of the design of experiments" and "analyse, interpret and draw inferences from results". Clearly these are two skills which lend themselves particularly well to this approach since even if these skills were being assessed within the context of a piece of practical work, pupils would probably have their written records of the work marked to assess these skills. This may, in fact, be the most useful way to use this technique to assess any of the skills for which it is used since this ensures that the assessment takes place within the context of a practical exercise.

The approach does have its detractors, however. Many teachers would recognise that by divorcing assessment from an actual practical exercise the results are less likely to reflect pupils' true level of attainment in the more practical skills. The APU (1985) have reached the same conclusions about the use of such "paper and pencil" analogues. It should also be remembered that the National Criteria for Biology state that written papers testing practical skills can supple-ment but not replace laboratory based assessment of practical skills.

Fig. 3.1 gives examples of questions requiring written answers which could be used to assess the objectives for which this strategy is appropriate.

Demonstrations

The increasing use of student practical work has led to a decline in the use of teacher demonstrations in schools. Nevertheless a case can be made for using demonstrations as the basis of some practical assessment. An experiment could be demonstrated to the class and the teacher could talk through the observations and/or measurements that can be made. The pupils might then be assessed on their ability to record accurately and clearly the results of the experiment (Objective 5). The demands made on resources and technical support is low, the pupils are working from a common set of experiences and the class can be more easily controlled than during practical work.

The need to ensure that one skill does not interfere with the performance of the assessed skill is particularly important to bear in mind when assessing "the ability to interpret results". For example, if pupils are expected to conclude from an observation of the movement of coloured liquid in an osmometer, that the solution in the beaker is less concentrated than the solution inside the selectively permeable membrane of the osmometer then it is important that the observation has been correctly made so that the attempt to draw a conclusion is based on the same observation for every pupil. This management strategy is useful here because the teacher, using a large demonstration osmometer, could ensure that all pupils made the same observation.

A useful modification of this strategy is to use it to assess a proportion of the class one at a time by setting up a demonstration which pupils take turns to look at during the course of a lesson. This assumes, of course, that the class is otherwise engaged in some other activity.

The advantages of this strategy, therefore, are clear. The most important disadvantage is true of demonstrations in general; that is that the more pupils participate in their practical work, the more they are likely to understand and learn from it. In terms of the assessment of practical skills the disadvantage that a distorted view of pupils' level of attainment will be gained using this strategy can also be argued.

Fig. 3.2 gives an example of an exercise which involves a demonstration that could be used to assess the ability to observe.

Stations

The simplest strategy which actually involves students doing the exercise on which the assessment is based, is to organise a number of "stations" in the laboratory. Pupils move from one station to another and at each one carry out a task. The individual exercises are short and usually all of roughly the same length. It would be possible to use this approach for all but two of the assessment objectives. "Following instructions" and

Fig. 3.1 Examples of questions requiring written answers.

Objective 2: Handling apparatus and materials

Put the following instructions for setting up a microscope to observe a specimen on a slide in the correct order:

1 Adjust the mirror so that light is shining through the specimen.

2 Whilst looking down through the eyepiece, turn the focussing knob so that the objective lens moves away from the stage to focus on the specimen.

3 Place the slide holding the specimen on the stage and hold it under the clips.

4 Place the microscope in a well lit area, eg in front of a window.

5 Whilst watching the objective lens move down, turn the focussing knob so that the objective lens moves down towards the stage but make sure it does not touch the slide.

Objective 3: Observation

The diagrams below show two Arthropods.

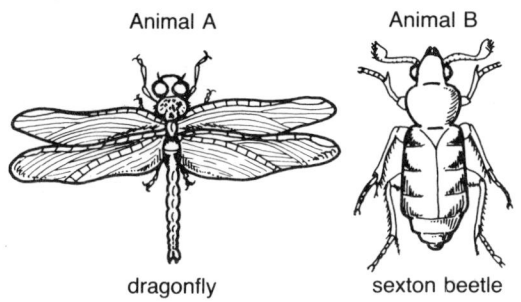

Animal A — dragonfly

Animal B — sexton beetle

Write down any similarities that you can observe between the two Arthropods.

Objective 4: Measurement

What is the volume of liquid in this measuring cylinder?
_____ cm³

Objective 5: Recording

Some pupils carried out an experiment to find out whether blowfly larvae preferred to live in light or dark conditions. They set up a choice chamber so that one side of it was dark and the other was light. They placed 20 larvae in the middle of the choice chamber and recorded how many larvae were in the dark and how many in the light, every minute. They found that after 1 minute there were 10 in the light and 10 in the dark. A minute later there were 12 in the dark and 8 in the light and a minute after that there were 14 in the dark and 6 in the light. After 4 minutes, however, there were only 12 in the dark and 8 in the light, but a minute later there were 16 in the dark and only 4 in the light. After 6 minutes there were 19 in the dark and one in the light. After 7 minutes there were still 19 in the dark and one in the light but after 8 minutes there were only 17 in the dark and 3 in the light. After another minute there were 18 in the dark and 2 in the light and a minute later, when the experiment ended, there were 19 in the dark and one in the light.

Draw a table of the results of the experiment.

Objective 6: Interpretation of results

A student found that an enzyme worked best at pH 2. From this result, choose which of these conclusions is a fair one:

A the enzyme works best in slightly acidic conditions
B the enzyme works best in slightly alkaline conditions
C the enzyme works best in strongly acidic conditions
D the enzyme works best in strongly alkaline conditions

Objective 7: Designing an experiment to test a hypothesis

A student was investigating the effect of surface area on the speed that acid diffuses into the centre of cubes of jelly. He predicted that:

"The smaller the cube, the quicker the acid will reach the centre of the cube".

Describe an experiment you could do to test if this hypothesis is correct.

Fig. 3.2 Example of a demonstration assessment exercise.

Objective 3: Observation

HEART STRUCTURE

Instructions to pupils:
Your teacher has placed a heart which has been cut open on the front bench. At some point she/he will call you up to examine the heart. When you are called up to examine it, follow these two instructions:

1 Study the section of the heart and write down any major features you observe.

2 Your teacher will show you the left and right ventricles. Study them both carefully and write down any differences you observe between them.

"Designing an experiment to test a hypothesis", would both take too long to be assessed using this strategy.

With small teaching groups it might be appropriate to have all of the students working on the assessment at the same time. Choosing activities which can be marked by inspecting students' written work obviously makes this more feasible.

For a larger class of, say, thirty students it would be tempting to set up thirty stations of ten different types. However neither the equipment or the time required to set up so many stations, is likely to be available. Even if these problems could be overcome if some of the stations involved the immediate checking of students' work, then the amount of marking to be done is likely to make this approach unmanageable. For such large teaching sets, a more realistic strategy would be to use a limited number of stations as one part of a lesson which also includes some other activity for the students.

This would be especially attractive if it was possible to design the assessment to be carried over two lessons by planning other activities which require two lessons to complete. Ideally, this other activity would involve the students in individual study and would occupy them so fully that the teacher would be released to supervise the stations. If the assessment was to be carried out in one lesson then the most obvious solution would be to link the practical assessment with a written test. Alternatively the teacher might aim to assess fewer pupils.

As with the previous two strategies, the main drawback of this approach is the artificial situation it creates. By concentrating on separate skills in very short exercises it discourages pupils from appreciating the overall process involved in the "scientific way of working".

Fig. 3.3 gives some examples of exercises which use the stations approach to assess objectives for which this strategy is appropriate.

Exercises which leave a record

Most of the routine practical work that students undertake on a biology course involves them making a written record of what they have done. Observations, conclusions, measurements, interpretations of results and experimental designs, can be recorded and the marks awarded used to produce an assessment grade. A number of the specimen exercises in Chapter 5 require this kind of written record to be made.

This approach of leaving a record of the work done can also be used for assessing such things as "following instructions" and "manipulating apparatus and materials" because the record need not be a written one. It is

possible to design exercises which involve pupils in producing work that allows assessment of several different objectives in one exercise. Furthermore, it is feasible to give such an exercise an overall purpose. For example, an exercise in which pupils make temporary mounts of biological material for microscopic examination can be used to assess "handling apparatus and materials" by examining how well the pupil has made the slide after the exercise is complete. Pupils can leave their mounts intact when they have examined them and these can be assessed at a convenient time. The ability to record using diagrams can also be assessed after the exercise by marking the diagrams drawn during the practical.

The main advantages of this technique are that:

(a) the teacher is left free from the task of assessment during the lesson since the records can be marked after the lesson if necessary
(b) the exercise can be part of the normal scheme of practical work in the course so that the assessment is taking place within the context of normal laboratory work.

The main disadvantage with this strategy is the problem that has to be faced in using any strategy which involves each pupil in practical work during the assessment exercise, viz the demand on apparatus and space is great since each pupil to be assessed must work alone. Again, this can be alleviated to some extent at least by using a short assessment exercise so that it occupies only part of the work to be covered during a lesson, or pair of lessons.

Fig. 3.4 gives examples of exercises which leave a record for assessment.

Fig. 3.3   Examples of station assessment exercises.

Objective 2: Handling apparatus and materials

Pupils are provided with a clamp stand, boss and clamp and a test tube at a station.

Set up the clamp stand, boss and clamp so that the test-tube is held in the clamp.
WHEN YOU HAVE FINISHED, PUT UP YOUR HAND SO THAT YOUR TEACHER WILL COME AND SEE WHAT YOU HAVE DONE.

Objective 3: Observation

Pupils are provided with 5 test-tubes, a test-tube rack, 5 solutions of different pH labelled 1 to 5, and universal indicator paper.

1 Put the five test-tubes into the test-tube rack.

2 Pour 1 cm depth of solution 1 into the first test-tube, 1 cm depth of solution 2 into the second test-tube and so on until each test-tube has 1 cm depth of a different solution in it.

3 Add 5 drops of universal indicator solution to test-tube 1. Write down the colour that the solution changes to.

4 Add 5 drops of universal indicator solution to test-tube 2. Write down the colour that the solution changes to.

5 Add 5 drops of universal indicator solution to test-tube 3. Write down the colour that the solution changes to.

6 Add 5 drops of universal indicator solution to test-tube 4. Write down the colour that the solution changes to.

7 Add 5 drops of universal indicator solution to test-tube 5. Write down the colour that the solution changes to.

8 Wash out all of the test-tubes.

9 Put all of the apparatus where you found it when you started work at this station.

Provided that the colour change for each solution is determined beforehand by the teacher, then examination of each pupil's record will enable assessment of the objective of observation.

Objective 4: Measurement

1 What is the volume of the liquid in the measuring cylinder?

2 What is the temperature of the water in the beaker?

3 What is the reading on the stopclock?

4 What is the mass of seeds on the balance?

5 How long is the leaf?

Objective 5: Recording

Pupils are provided with 10–20 dishes, each containing a number of beans/peas/other large seeds.

1 Count the number of beans in each of the dishes at the station. Record your results in a table.

Fig. 3.4 Examples of exercises which leave a record.

Objective 2: Handling apparatus and materials

In this exercise pupils make a temporary mount of onion epidermal cells and examine it microscopically. The mounts should be labelled with pupils' names and collected in for assessment after the lesson, or they can be kept by pupils during the lesson and assessed when it is convenient to do so during the lesson.

LOOKING AT PLANT CELLS

Instructions to pupils

In this exercise you are going to make a temporary mount of some onion cells and examine them with a microscope so that you can see something of their structure.

1 Use a scalpel to cut out a small piece of the thin membrane that can be found between the layers of an onion. A piece about 0.5cm × 0.5cm is large enough.

2 Remove the piece of membrane from the onion using forceps and place it on a microscope slide.

3 Add a drop of iodine solution to stain the cells.

4 Carefully lower a coverslip onto the tissue.

5 Set up the microscope so that you can examine the cells on the slide under the highest power possible.

6 Record your observations of ONE cell in the form of a drawing.

7 Keep your slide with you when you have finished so that your teacher can examine it.

Objective 3: Observation

In the course of attempting to build up a picture of the food web in a soil pupils remove animals from a soil sample which they have collected and try to identify them using a key in this exercise. Pupils should place an animal they have found in a small dish (such as a petri dish) and label this with the name of the animal from the key and their name. The dishes can be collected in for assessment later or assessed at a convenient point during the lesson.

Instructions to pupils

1 Collect a sample of soil from the habitat you are interested in. About half a washing up bowl full is enough.

2 In the laboratory spread out several sheets of newspaper on the bench and empty the soil onto this.

3 Using a blunt instrument such as a spatula or seeker, very carefully sift through the soil looking for any animals.

4 When you have found an animal, place it in a dish.

5 Use the key to soil animals to identify the animal.

6 When you have identified the animal, write its name on a label and write your name on the same label. Stick the label onto the dish and keep this for your teacher to see.

7 Keep all the animals you have removed until the end of the lesson. Then they will be put back where you found them.

Objective 6: Interpreting results.

HOW EFFECTIVE IS SWEATING AT COOLING YOU DOWN?

Instructions to pupils

1 Set up the apparatus as shown in the diagram below. MAKE SURE THAT THE PAPER AROUND EACH TEST TUBE IS THE SAME THICKNESS.

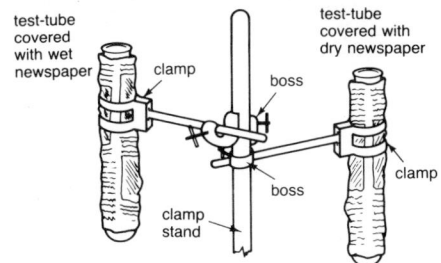

2 Copy down the results table:

Time (minutes)	Temperature (°C)	
	Dry tube	Wet tube

3 Fill both test-tubes to the top with hot water from the same source, so that the temperature is the same in both. MAKE SURE THAT THE WATER IN BOTH TUBES IS AT THE SAME TEMPERATURE.

4 Measure the temperature of the water in each of the tubes and record these in the results table at time 0 minutes.

5 Measure the temperature of the water in each tube at intervals of one minute for the next 15 minutes and continue to record these in the results table.

MAKE SURE THAT YOU GET WHAT EVER ASSIST-ANCE YOU MAY NEED TO MEASURE THE TEM-PERATURES OF THE WATER IN THE TUBES AND/OR TO RECORD THE RESULTS IN THE TABLE.

6 When you have finished your experiment, plot your results on a graph with the axes shown below. Put both lines on the same graph.

MAKE SURE THAT YOU GET WHAT EVER ASSIST-ANCE YOU MAY NEED TO PLOT YOUR GRAPH.

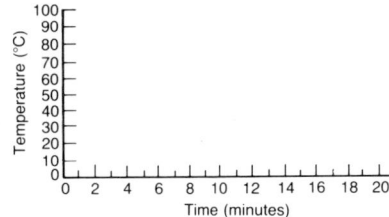

7 From your table, what was the temperature of the water in the dry tube at 8 minutes?

8 From your graph, what was the temperature of the water after 4½ minutes?

9 Describe what pattern you see in the results for the wet tube.

10 How does the pattern of the results for the dry tube compare with the results for the wet tube?

11 What conclusion do you draw from these results?

12 Explain why there is a difference between the water in the tubes in how quickly the water cools down.

13 What would the temperature be in the dry tube after 17 minutes?

Using an additional assessor

The assessment of many of the objectives would ideally be carried out by the teacher actually watching the pupil doing the task. This poses substantial problems for a teacher trying to teach and supervise the class in safe practical work at the same time as assess pupils. Whilst it is appreciated that this strategy will be nearly or completely impossible to use in many departments, using an additional assessor is one strategy which enables the teacher to satisfactorily assess a whole class by ensuring that the class is being supervised even if the teacher doing the assessment is occupied with only one pupil. This strategy becomes especially necessary when there are pupils in the class who are likely to be disruptive to the smooth execution of the practical work if they are not closely supervised.

The second teacher could have one of two roles:

1 Take responsibility for assessing half of the class, thereby reducing the time needed for the assessment of the whole class, or

2 Act as the class teacher, leaving the actual teacher free to do the assessment, or vice versa.

Some schools may be able to provide cover for the class of one biology teacher who is then released to assist another colleague in such an assessment. This is especially possible when there three or more small classes being taught at the same time in a science department since one member of staff could teach two small groups together. An alternative would be to ask for time to carry out an assessment of one or more skills during the trial examination period.

Chapter 5 includes 3 examples of assessments designed specifically for use with this strategy, but clearly almost any practical exercise for which there is sufficient equipment and preparation time for the whole class to be involved could be used for assessment using an additional assessor.

Fig. 3.5 gives an example of an exercise which uses an additional assessor.

Fig. 3.5   Exericise which makes use of an additional assessor.

(See Chapter 5, Exercise 5)

WHAT IS THE BEST pH FOR PEPSIN TO WORK?

Introduction

Pepsin is an enzyme which digests protein. It is found in the stomach, therefore it is likely that the pH in the stomach is the pH for pepsin to work best. This experiment is designed to test this hypothesis:

The pH in which pepsin works best is the same as the pH of the stomach.

The pH in the stomach is acidic. You are going to set up mixtures of protein and pepsin at different pH's to find out whether or not pepsin digests the protein most quickly in acid conditions.

Instructions

1 Set up a water bath at 37°C and keep it at this temperature throughout the experiment.

2 Place 6 test-tubes in a rack and number them 1 to 6 as shown below:

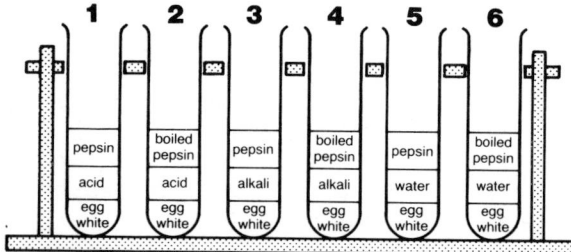

YOU ARE NOW GOING TO MEASURE OUT SEVERAL SOLUTIONS INTO THE TEST-TUBES AS SHOWN IN THE DIAGRAM ABOVE. WHILE YOU ARE DOING THIS YOUR TEACHER WILL WANT TO ASSESS YOUR ABILITY TO MEASURE WITH A METRIC RULER.

3 Pour 1cm depth of egg white suspension into each of the test-tubes.

4 Pour 1cm depth of acid acid into two of the test-tubes; 1cm depth of alkali into two of the test-tubes and 1cm depth of water into two of the test-tubes.

5 Pour 1cm depth of pepsin solution into one of the tubes containing acid, one of the tubes containing alkali and one of the tubes containing water.

6 Pour 1cm depth of boiled pepsin solution into each of the three remaining test-tubes. These tubes are controls for the tubes containing pepsin to make sure that the protein is only digested when you put enzyme into it.

7 Shake the tubes well to mix the contents.

DO NOT GO ANY FURTHER UNTIL YOUR TEACHER HAS CHECKED YOUR APPARATUS

8 Place the six tubes in the water bath and start the stopclock.

9 Record the time it takes for the egg white suspension to go clear in each of the tubes in a table like the one below. Copy out this table while you are waiting for the pepsin in the tubes to go clear.

WHILE YOU ARE DOING THIS YOUR TEACHER WILL ASK YOU TO MEASURE THE TEMPERATURE OF THE WATER WITH THE THERMOMETER AND ALSO TO MEASURE THE TIME ON THE STOPCLOCK.

Tube Number	Conditions	Time taken for suspension to clear
1	Acid	
2	Acid control	
3	Alkali	
4	Alkali control	
5	Neutral	
6	Neutral	

10 Examine the results in your table. Which pH conditions does it appear that pepsin works in best? Is it the the same pH as the stomach? If so, this means that the hypothesis must be correct.

Assessment during routine practical coursework

A number of the examining groups make it clear that not all candidates need to be assessed at any particular time or on the same exercise. This is a most important provision and would certainly make it easier to use any of the strategies described above. It also opens up the possibility of carrying out small–scale assessment of pupils during routine coursework. Many teachers will find this the most attractive strategy to use since it ensures both that the pupils are active in learning through their practical work, regardless of whether assessment of skills is being carried out, and because it reduces the pressure on teachers to set up special circumstances under which assessment can take place. In addition it allows for the assessment of pupils who are absent on the occasion of the assessment.

If the approach was being used for the assessment of manipulative, measurement or recording skills, the teacher could identify a number of skills which are likely to be used over a certain period – say half a term. Using the kind of grid illustrated below, the performance of a number of pupils on one or more skills could be recorded in a particular lesson. The actual number of pupils assessed or the actual number of manipulative, measurement or recording skills which are assessed on this occasion does not matter. By using a grid system those pupils who have not as yet been assessed can be identified and the teacher can aim to assess some of them on the next occasion that it becomes convenient to do so. The key is that the teacher has identified a sufficiently large set of occasions on which assessment can occur to ensure that all pupils are assessed but that the assessment does not hinder the normal teaching of coursework through practical experience.

Having drawn up the performance criteria for each level of attainment for each assessment objective, the grid would contain all of the skills to be assessed within each objective, ie apparatus which will be used to assess manipulative skills or instruments to be used for measurement etc. Using the example of assessment of the skill "manipulate apparatus and materials", the pupils would then be assessed over a period of time on apparatus in the list shown on the grid. If the performance criteria outlined in Chapter 2 for this skill were used, a pupil who showed ability to successfully manipulate any of the apparatus in the grid would gain a high level of achievement (e.g. an "L" on the SEG scheme, or a "9" on the MEG scheme). A pupil who showed ability to manipulate any of the apparatus, except that designated as needing a high level of attainment to use (marked ● in the grid shown) would gain a medium level of attainment, and pupils able to manipulate "every day" apparatus (marked ○ in the grid shown) would gain a low level of attainment. The successful use of a piece of apparatus could be recorded with a tick.

In using the grid system it is not essential that pupils are assessed on all of the apparatus in the list, but a sufficiently large range of pieces of apparatus so that the teacher feels able to make an objective judgement of the pupil's level of attainment. Thus the grid represents a profile of pupils' performance on a range of tasks. The system is sufficiently flexible, however, so that it could be used to assess all of the tasks instead of just a range of them if the teacher so desires.

In practice, what ever management strategies are preferred, it is unlikely that all of the individual skills listed in the performance criteria of the objectives "handlng apparatus and materials", "observation", "measurement" "recording" and "interpretation of results" will ever be assessed using the same exercise. This being true , it is therefore necessary to have some form of "master grid" in which marks can be recorded for pupils' attainment in each of the individual skills within the objectives. A master grid of this nature has been used extensively in Chapter 5. Fig. 3.6 is an example of such a grid.

Fig 3.6   An example of a grid used in assessment of practical skills during routine coursework.

Objective 2: Manipulate apparatus and materials										
Apparatus / **Name**	Thermometer°	Clamp stand boss and clamp	Potometer●	Microscope	Forceps°	Scalpel°	Temporary mount●	Water bath		
A. Armitage	✓				✓	✓				
J. Bean	✓	✓		✓	✓	✓		✓		
L. D'Arcy	✓	✓	✓	✓	✓	✓	✓	✓		
M. Everett	✓	✓		✓	✓	✓	✓	✓		
J. Jones	✓			✓						
C. Malone	✓	✓	✓	✓	✓	✓	✓	✓		
T. Turner	✓	✓		✓	✓	✓		✓		

KEY
● = apparatus requiring a high level of manipulative ability
○ = 'every day' apparatus requiring a low level of manipulative ability.

Summary Table

It is clear from the discussion in this chapter that although many of the management strategies can be used to assess a wide range of assessment objectives, some strategies are more suitable for use in assessing some objectives than others. Fig. 3.7 is a Summary Table showing which management strategies are most suitable for use for assessment of each objective.

Fig 3.7 Summary Table showing Management Strategies and Assessment Objectives for which they are most appropriate.

① = Not wholly suitable
② = Not necessary in view of difficulty of arranging an additional assessor

Management strategies \ Assessment objectives	Following instructions	Handling apparatus and materials	Observation	Measurement	Recording results	Interpreting results	Designing an expt. to test a hypothesis
Written answers		①√	√	√	①√	√	√
Demonstrations			√		√	√	√
Stations	①√	√	√	√	√	√	
Leaving a record		√	√	√	√	√	√
Additional assessor	√	√	√	√	②√	②√	②√
Routine coursework	√	√	√	√	√	√	①√

4 Designing assessment exercises

Introduction

Now that consideration has been given to drawing up lists of skills within each assessment objective, defining performance criteria for each objective and the management strategies that are available, a teacher or department is in a position to develop a scheme of exercises which can be used for assessment purposes.

Developing practical skills

It seems obvious that if pupils are to have their competence in a range of skills assessed then they must be given ample opportunity to develop those skills prior to the assessment. Notwithstanding this necessity, it is quite possible that unless a scheme of development and assessment of objectives is drawn up for the whole course, the pressure associated with teaching GCSE syllabi may lead to inadequate development of skills. To avoid this and to ensure that all teachers feel secure in the knowledge that sufficient opportunities for development and assessment of skills are provided during the course it is important to review the departmental Teaching Scheme and the suggested practical exercises associated with it, to establish exactly where each skill can be developed and then assessed. The Teaching Scheme should state clearly which skills are to be developed and/or assessed during each practical exercise. Fig. 4.1 illustrates how this has been done for part of the Teaching Scheme used to teach the NEA Biology syllabus in one school. The matrix contains the titles of the practical exercises which are used during the course of teaching each area of the syllabus down one axis and the skills which will be developed and/or assessed during each exercise along the other axis.

Differentiation

The General Criteria state (p.2) that:

"All examinations must be designed in such a way as to ensure proper discrimination so that candidates across the ability range are given opportunities to demonstrate their knowledge, abilities and achievements: that is, to show what they know, understand and can do."

All assessment exercises must also be designed to ensure that they adequately differentiate across the attainment range. Thus candidates who are capable of carrying out less demanding tasks must be presented with tasks which provide opportunities for them to demonstrate these skills. At the same time higher attaining candidates must be given the chance to demonstrate their command of more sophisticated skills and discrimination between candidates must be possible using the exercise.

Differentiation can be achieved in three broad ways: aiming an activity at a particular attainment level, using stepped exercises and using stepped markschemes.

(a) Exercises specifically aimed at an attainment level

Reference to the performance criteria suggested in Chapter 2 ensures that aiming exercises at particular attainment levels is straightforward. If the objective "recording results" is considered, then a relatively low attaining pupil would be set the task of recording the results from an experiment in a table with which he/she has been provided. Thus the pupil can be successful in recording the results without being presented with the more demanding task of first drawing up the table. A higher attaining pupil might be set the task of drawing the table before filling it in.

Devising such exercises seems more straight forward than using stepped exercises, but in their purest form they can only be used if all of the pupils within a group are at the same level of attainment. Even if tight streaming is employed this is unlikely to be the case. Another major disadvantage of such exercises is that the teacher pre–judges the likely attainment of candidates and therefore does not allow the candidate the opportunity to be successful in more demanding tasks or to demonstrate a higher level of achievement in the objective being assessed. Although teachers may feel that they know their pupils' abilities well enough to make such judgements, this cannot be guaranteed.

(b) Stepped exercises

Where there is any spread of attainment in a class (as there almost always will be), a stepped exercise will be necessary if the exercise is to enable all candidates to demonstrate what they can do.

Stepping an exercise involves arranging the tasks within the exercise into ascending order of difficulty. Take, for instance, an exercise in which an experiment to find out what effect temperature has on the activity of amylase is to be carried out, and in which the assessment objectives are to measure with a thermometer, measure out a liquid with a measuring cylinder and measure out a mass of dried enzyme using a top-pan balance. The pupils would be presented with these tasks in the order in which they are written here since this is likely to be the order of difficulty of the tasks.

(c) Stepped markschemes

Stepped markschemes are useful in assessing objectives where all pupils can demonstrate some level of achievement, even if this is at a low level, because the task is sufficiently open–ended to allow this. Assessment of the ability to design experiments, for example, lends itself to this approach. Provided the teacher ensures that pupils understand the task to be carried out, all pupils can show some ability to design an experiment and are given the freedom to do what ever they can. The markscheme is devised to give credit for particular features in the design, the more are present, the greater the level of success. "Recording results" and "interpreting results" are other objectives which also lend themselves to this approach.

A disadvantage of stepped exercises, however, is that an increasing order of difficulty may not be the

Fig 4.1   Example of matrix of practical exercises and assessment objectives for part of a GCSE Biology Teaching Scheme

Group headers across the numbered columns:
- MEASUREMENT: 1(a b c), 2(a b), 3(a b), 4, 5
- OBSERVATION: 6, 7, 8, 9, 10
- HANDLING: 11, 12, 13, 14
- RECORDING: 15, 16, 17, 18, 19
- INTERPRETATION: 20(a b), 21, 22(a b), 23, 24, 25, 26
- DESIGN: 27, 28, 29, 30, 31, 32

G.C.S.E. Biology Experiments and Skills	1a	1b	1c	2a	2b	3a	3b	4	5	6	7	8	9	10	11	12	13	14	15	16	17	18	19	20a	20b	21	22a	22b	23	24	25	26	27	28	29	30	31	32
Diversity of life																																						
1 Investigating variation circus	1	1	1	1	1	1	1	1		1			1						1	1	1	1	1			1	1	1			1				1			
2 Taxonomic groups											2		✓																									
3 Using Keys to identify organisms									✓					✓																								
Ecology																																						
1 Finding the amount of energy in a plant community																																						
2 Making a pyramid of biomass for a soil community						2														2	✓			2							2				2			
3 Effect of light on ground flora									✓																													
Cell biology																																						
1 Microscopic examination of cells								2		3	✓																											
2 Microscopic examination of tissues																	✓	1																				
Biological molecules																																						
1 Using food tests to identify mystery solutions	2																													✓						1		
2 Effect of temperature and pH on amylase activity	2															✓					✓								✓									
3 Effect of enzyme and substrate concentrations on pepsin activity																						2				✓			1									
4 Removal of stains using biological and non-biological washing powders						3											1																					
Respiration																																						
1 Production of carbon dioxide during respiration																																	✓					
2 Effect of temperature of incubation on bread raising																					2					2								✓				
3 Brewing								1	3									1																				
Plant nutrition																																						
1 Effect of bicarbonate ion concentration on rate of photosynthesis															✓	1																						
2 Effect of light intensity on rate of PSN																															1							
3 Need for mineral ions	2																		✓																			
4 Leaf structure in relation to function in PSN													1	✓																✓		✓						
Gaseous exchange in plants and diffusion																																						
1 Effect of SA and SA/VOL ratio on diffusion	3																						2	3		3												
2 Microscopic examination of leaf structure vs. function in gaseous exchange														1					1																			

✓ = opportunity to develop this skill
1 = 1st assessment of skill
2 = 2nd assessment of skill
3 = 3rd assessment of skill

most logical order in which tasks should be carried out in an exercise.

Notwithstanding the usefulness of stepped exercises and mark schemes in mixed attainment classes, it is unreasonable to expect an exercise to differentiate across the whole attainment range. Also, classes will often contain a restricted range of attainment. For both these reasons it may be necessary to devise exercises which are aimed at a smaller range of attainment. In order to devise an exercise which can successfully differentiate and discriminate, it is important to have a clear understanding of which tasks are difficult and which are easier. This information is required in order to draw up the performance criteria for each objective. Several sources of information exist which can guide teachers into assembling a rank order of difficulty. These include the grade descriptions provided by the examining groups themselves and the draft Grade Criteria provided by the Biology Working Party of the Secondary Examinations Council.

Isolating the assessment objectives

It follows that if a sequence of tasks is to be carried out during an exercise in which only one or two of them are to be objectives for assessment, then the other tasks within the exercise must not prejudice a pupil's chance of demonstrating competence in the objective(s) to be assessed. Thus if the successful completion of the objective to be assessed is contingent upon successful completion of some previous task (whether the previous task is to be assessed or not), then the teacher must ensure that the previous task has been successfully completed before attempting to assess the latter objective. It is advisable to ensure that tasks to be carried out prior to the assessed task are sufficiently simple to permit pupils to carry them out successfully.

An example of this situation is where a candidate may be required to manipulate apparatus to carry out an experiment in which the ability to record results will be assessed. It is clearly important that the ability of pupils to record the results is not affected by how successfully pupils assemble the apparatus and carry out the experiment. The teacher would be required to provide such help as might be necessary to ensure that these previous tasks were successfully carried out before attempting to assess the ability to record results.

Reducing factual and concept load

Evidence from the APU suggests that the form and context in which a task is presented can affect the ability of the candidate to carry it out. It is therefore important that, when devising an exercise to assess an objective, teachers ensure that the factual information and/or the concept load involved in understanding the purpose of the task does not prejudice a candidate's chance of success.

An example is an experiment to determine how the size of a cube of gelatin affects how quickly it is digested by an enzyme. Knowledge of the term "protease" is not essential to a candidate's ability to manipulate apparatus or measure how long it takes for the cube to dissolve, for instance. Further, when interpreting the results, candidates could correctly state that the smaller the cube, the quicker it dissolves without the need for understanding that the protein in the gelatin has been digested to soluble amino acids which have then dissolved. Burdening candidates with the need to understand in detail what has happened to the gelatin could prejudice their ability to state what effect the size of the cube has on the rate at which an enzyme digests it.

Language

In addition to ensuring that the factual and concept loads of assessment exercises do not prejudice candidates' chance of success, the language level of instructions, whether written or oral, must be sufficiently simple to ensure that all candidates can fully understand the task they are to perform. Further, any candidates who experience difficulty in reading or understanding instructions must be given as much help as necessary to enable them to attempt the task since their ability to comprehend instructions is not an assessment objective. Where possible, diagrams in conjunction with instructions are desirable.

Simplicity of exercises

For all the reasons outlined above, and the practical difficulties of equipment shortages, lack of technician support, lack of preparation time and so on, it seems sensible to ensure that any exercises devised for the school-based assessment of experimental work are kept as simple as possible. Complex experimental work may have a place in teaching the syllabus content and developing skills but places great strain on teachers and technical staff if the criteria for devising suitable exercises outlined in this chapter are to be satisfied.

Grading

Once the performance criteria have been drawn up for each of the assessment objectives in the scheme being used by a biology department, it is necessary to find a way of putting these criteria into practice during actual practical exercises. It is unlikely that all of the skills in the list drawn up by the department for the performance criteria of any of the objectives will be assessed during a single exercise. What is needed is a way of recording competence in each skill within an objective and then awarding a grade in accordance with the performance criteria.

A straightforward approach to this would be to draw up a list of marking points for each of the skills within each objective when ever that skill was to be assessed. In the example of "handling apparatus and materials" each piece of apparatus would have a list such as that below for a clamp stand, boss and clamp:

- boss attached to clamp stand correct way up
- boss adequately tightened onto clamp stand
- clamp attached to boss tightly
- clamp positioned so that weight to be held is over base of clamp stand.

These marking points would be used as a checklist during the assessment of this skill. If a pupil has done all of the things in the list, a record would be made on a master grid such as that shown in Figure 3.6 that he/she is able to use this piece of apparatus. This would be done for the assessment of all of the pieces of apparatus in the list which would appear on the master grid for this objective. At the end of the course each pupils' achievements would be used to produce a grade in accordance with the performance criteria. Thus if a pupil was able to use all of the items in the list except those marked "●" then he/she would achieve a grade indicating a medium level of attainment.

A modification of this method would be to attempt to assess only a representative sample of the items in the list for each objective which, of course, has the advantage of reducing the time which has to be devoted to assessment and allows more time to be spent teaching.

It would be possible to use the method without having a checklist for each item in the list of skills; provided a teacher has a clear understanding of what has been agreed as competence in carrying out a skill then a pupil's competence can be recorded in the master grid much more quickly without first going through the exercise of marking against the checklist.

Summary

Fig. 4.2 shows a summary flow chart to illustrate the sequence of processes involved in producing a scheme of internally assessed practical work. The sequence begins by drawing up a scheme of practical work for the whole GCSE Biology course which could be used to teach the syllabus. The intention, therefore, is very

much that the teacher or department draws up a scheme of internal assessment of practical skills which fits the scheme of practical work being used to teach the syllabus, rather than producing a scheme of practical work used to assess skills around which the teaching of the syllabus must fit.

Fig 4.2   Summary Flow Chart for drawing up a scheme of internally assessed practical work

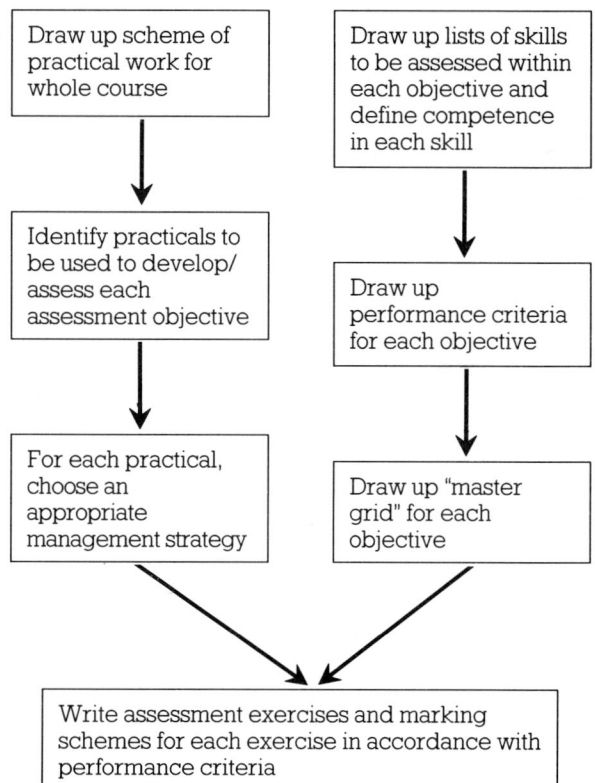

5 Specimen assessments and teaching notes

Introduction

This chapter provides 20 exemplar exercises intended to illustrate how the objectives outlined in Chapter 2 can be assessed using the management strategies suggested in Chapter 3. All of the exercises are designed to assess part or all of one assessment objective. Where only part of an assessment objective is assessed, the exercise is used to assess one or more of the skills which have been suggested for that objective in the performance criteria drawn up earlier. Thus the exercise provides an assessment of only part of the overall objective. In order to record the attainment of any of these skills within an objective it is suggested that a "master grid" is drawn up for each assessment objective containing the names of the pupils and the skills for that objective as illustrated in Fig. 3.1. When a pupil achieves competence in a skill within an objective a "tick" is made against that pupil's name for that skill.

In choosing which exercises should be included, an attempt was made to satisfy the following criteria.

1 The exercises should cover as complete a range of subject material as possible so that assessments can be made at any stage of a course and in the context of any topic.

2 In attempting to achieve differentiation, as noted in Chapter 4, it is undesirable to construct exercises which differentiate across the whole range of attainment. Thus some exercises are included which are intended for use with the top half of the attainment range, some with the bottom half and only a small number for use with any part of the attainment range. Some exercises are suggested for use with smaller ranges of attainment.

3 Exercises should make reasonable demands on the equipment of every biology department. To this end all of the exercises use only simple equipment which most departments would be expected to have. No exercises have been included which require expensive or specialist equipment. This does not mean, of course, that equipment of this nature could not be used for assessment purposes, particularly in the context of a demonstration.

4 Exercises should be familiar to teachers so that they can have confidence both in the setting up and the outcome of the exercises. Other exercises should include technological and sociological applications of biology in line with the requirements of the National Criteria. Where exercises have been aimed at teaching these applications, the practical work itself is familiar; it is the use to which that work is put which is new. It is not necessary to invent entirely novel practical exercises in order to teach such applications. Rather the use of well tried and tested exercises is likely to make the transition into the relatively unfamiliar area of teaching societal implications much easier. It is the teaching style which will ensure the success of such exercises: strategies such as investigational and problem solving approaches are to be encouraged at the expense of exercises which seek to merely illustrate theory with practical work.

This final criterion is a particularly important one for the exercises in this book to satisfy while teachers are attempting to satisfactorily cope with the new demands of developing and assessing skills which GCSE brings with it. Teachers must feel confident that the practical exercises themselves can be successfully used both for teaching and assessment and that they can be managed easily.

Although the teaching notes suggest how exercises could be used with particular management strategies, all of the exercises could be used with other management strategies. Equally, although it is suggested that particular assessment objectives are assessed using each exercise, many of the exercises could also be used to assess other objectives.

Index Table

Table 5.1. is an index table of the exercises included in this chapter. It shows which exercises illustrate the six management strategies and seven assessment objectives.

Table 5.1   Index table of exercises, assessment objectives and management strategies

MANAGEMENT STRATEGY	Following instructions	Handling apparatus and materials	Observation	Measurement	Recording	Interpreting results	Designing an experiment to test a hypothesis
1 Questions requiring written answers			3			8	13
2 Demonstrations			19			14	4
3 Stations	7			18	11		
4 Exercises which leave a record		15			2	12	6
5 Using an additional assessor	10	17		5			
6 Routine coursework	20	1	16	9			

Teacher's notes

Exercise 1

Assessment objective

Ability to handle apparatus and materials
(NEA Skill 11)

Previous experience

Knowledge and understanding of transpiration are required. In addition pupils must be aware of the sort of environmental factors which might effect transpiration because they have an affect on the rate of evaporation of water. The practical exercise assumes that the class has designed the experiment they are going to carry out to investigate the effect of changing factors in the environment on the rate of transpiration.

Management strategy

Assessment can be made during routine coursework.

Pupils design an experiment as a class to investigate the effect of envi. ˑmental conditions on transpiration rate. The experiment will probably have been designed in a previous lesson so that as much time as possible is available for this exercise. Each pair of pupils could then investigate the effect of changing one factor and the results from the whole class collated at the end to draw up the full list of factors.

The simple potometer is a very difficult piece of apparatus to assemble successfully, particularly filling the capillary tubing with water and preventing entry of air. For this reason a teacher could not rely entirely on written instructions and a diagram, however clear, to ensure success. To give pupils the best chance of assembling the apparatus correctly, it is important that the teacher first demonstrates how to set up the potometer. Particular attention should be given to showing pupils how to fill the capillary tubing with water and how the leafy shoot can be inserted without allowing air to enter. Clearly the success of the whole exercise can be dependent upon this. The instructions provided on the sheet then serve as a reminder to pupils to ensure that they are not being judged on there ability to remember instructions given by the teacher.

Differentiation is achieved in this exercise if pupils are given as much help as they require to assemble the apparatus and carry out the experiment. Notwithstanding this provision, the difficulty of the manipulation required in setting up the potometer dictates that this exercise is for the assessment of pupils with above-average attainment in this objective.

Equipment availability is likely to mean that pupils will work in pairs. The experiment therefore lends itself to assessment of this objective using the strategy of assessing just a proportion of the class during the experiment, and using the checklist of marking points below and recording competence in a 'master grid'.

The teacher would instruct one person from each pair of pupils to assemble the apparatus, preferably while the other member of the pair is engaged in other work (eg writing down an introduction and drawing out the results table for the experiment). The assessment can then take place at any time during the lesson. In this way, up to half the class can be assessed.

Since the abilities to record results and interpret results are not being assessed it is important that pupils are given as much help as they require in completing these parts of the exercise.

Time required   Approximately 1 hour

Resources required

Each of these items per pair of pupils
30 cm length of capillary tubing
30 cm ruler (alternatively a shorter length of capillary tubing and appropriately sized piece of metric ruler);
2 elastic bands
Plastic bowl or bucket (or access to sink with plug)
2 cm length of rubber tubing to fit tightly over capillary tubing
Leafy shoot of herbaceous plant (eg geranium);
Petroleum jelly
Stopclock or watch
Fan, plastic bag or heater (to provide changed environmental conditions);
100 cm^3 beaker
Clamp stand, boss and clamp.

Marking the assessment

The exercise can be used to assess the ability to handle two of the pieces of apparatus from the list which will have been drawn up in the performance criteria for this objective: assembling a clamp stand, boss and clamp, and assembling a simple potometer. Below are the suggested marking points for each of these skills. If a pupil has performed all of the manipulations in the marking list then he/she is awarded a 'tick' against his/her name for the piece of apparatus in the 'master grid'.

A teacher or department may choose to award a mark for competence if less than all of the marking points have been achieved, for instance if a pupil has achieved a majority of the points, or two thirds of the marking points, depending on how difficult manipulation of these pieces of equipment is thought to be.

Potometer:

– leaf surface not wet
– air tight seal of shoot around rubber tubing
– no mess in using petroleum jelly to seal potometer
– capillary tubing firmly attached to ruler
– potometer free from air bubbles except the bubble used for measuring rate of transpiration

Clamp stand, boss and clamp:

– boss attached to clamp stand correct way up
– boss adequately tightened onto stand
– clamp positioned so that weight of potometer is over base of clamp stand
– boss adequately tightened around clamp
– potometer positioned in clamp at appropriate height to be placed in beaker of water
– clamp placed at appropriate position around potometer to ensure secure support
– clamp adequately tightened around potometer

Exercise 1. Which Environmental Conditions Affect Transpiration?

Introduction

In this experiment you are going to find out what conditions around a plant affect how quickly transpiration takes place. To measure how quickly a plant is transpiring the equipment in the diagram can be used. This is called a **Potometer**.

Setting up the potometer

READ **ALL** OF THESE INSTRUCTIONS BEFORE STARTING

WHEN YOU HAVE READ THE INSTRUCTIONS, YOUR TEACHER WILL DEMONSTRATE HOW TO SET UP THE POTOMETER.

1 Assemble the apparatus you have been given to make a potometer as shown in the diagram below. While you are doing this make sure you do these things:

   (a) Keep the cut end of the shoot under water at all times.

   (b) Keep the glass tubing full of water while putting in the shoot.

   (c) Seal the join between the shoot and the glass tubing with petroleum jelly.

   (d) Do not get the leaves wet.

2 When it is assembled, clamp the apparatus with a clamp, boss and stand.

3 Allow a very small air bubble (about ½ cm long) into the open end of the glass tubing. This is done by taking the open end out of the small beaker of water and allowing air to be sucked up the tubing.

4 Put the open end of the glass tubing into the small beaker of water again.

IF YOU HAVE DIFFICULTY IN SETTING UP THE POTOMETER, ASK YOUR TEACHER FOR ASSISTANCE.

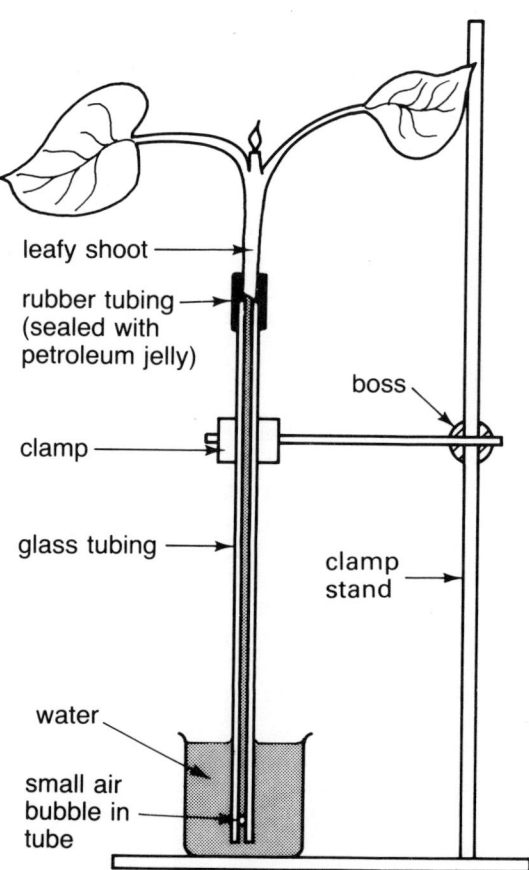

leafy shoot

rubber tubing
(sealed with
petroleum jelly)

boss

clamp

glass tubing

clamp
stand

water

small air
bubble in
tube

The experiment

There are two stages to the experiment:

(i)   Find out how quickly the shoot transpires in normal conditions.

(iii) Find out how quickly the shoot transpires with different conditions. The class will be investigating the effect of changing THREE factors around the plant. These are:

  (a) Increased temperature
  (b) Increased humidity
  (c) Increased air movement

Each group will investigate ONE of these factors. You will already know which condition you are going to change. You will also have decided how you are going to set up the new condition and how to keep everything else the same so that the test fair.

5 Copy the results table:

	Normal conditions	Conditions with one factor changed
Distance of air bubble at start		
Distance of air bubble at end		
Distance moved in time available		

6 Measure how far up the glass tubing the air bubble is at the start of the experiment.
   RECORD THIS IN THE RESULTS TABLE AS "DISTANCE OF AIR BUBBLE AT START".

7 Leave the shoot to transpire for about 15 minutes.

8 Measure how far up the glass tubing the air bubble is at the end of this time.
   RECORD THIS IN THE RESULTS TABLE AS "DISTANCE OF AIR BUBBLE AT END".

9 Take away the distance at the start from the distance at the end.
   RECORD THIS FIGURE IN THE RESULTS TABLE AS THE "DISTANCE MOVED IN TIME AVAILABLE".

10 Change the factor that you are investigating. Every other factor must stay the same.

11 Repeat stages 6, 7, 8 and 9 to complete the experiment.

Interpreting the results

12 In which conditions did the air bubble move most quickly?

13 Finally write a sentence to say what effect the factors you were investigating have on how quickly the plant transpires.

Teacher's notes

Exercise 2

Assessment objective

Ability to record results in diagrams
(NEA skill 15)

Previous experience

This exercise would probably come during a topic on
cells and tissues, and knowledge of cell structure will
be necessary to use the sheet as it is written here. In
addition, pupils will need the ability to use a microscope.

Management strategy

This is an exercise in which pupils leave a record. The
exercise is used to assess ability to record results in the
form of diagrams by instructing pupils to draw one cell
that they see on their slide and label it. The drawings
can then be collected in for assessment out of lesson
time.

It is unlikely that there will be enough microscopes
for every pupil to have one each and the exercise itself
has been kept reasonably short to take account of this.
It is therefore intended as an activity which forms part
of a lesson, pupils not engaged in the exercise being
set another activity. The exercise has been designed
for use across the whole attainment range. It is
important that pupils have a sufficiently good temporary
mount for them to see at least one onion epidermal cell
clearly enough to draw it. The teacher's major role
during the exercise, therefore, is ensuring that all
pupils have made their mounts and set up their
microscopes adequately. They should, of course, receive as much help in doing this as is necessary to
produce adequate preparations for viewing.

Activities for pupils to carry out when not being
assessed include identifying diagrams of cells from
different types of cells from different types of tissues
using other resources such as 35 mm slides, photographs or books. Alternatively, the second activity
could be a research exercise into the structures found
inside cells and the functions of these structures.

Time required   Approximately 1 hour

Resources required

Class set of microscopes
One or two onions
Forceps
Scalpels
Blunt seekers/mounted needles
Iodine in potassium iodide solution
Microscope slides
Coverslips
Filter paper or paper towels
Bench lamps

Marking the assessment

Using this strategy, a teacher can mark the drawings
out of lesson time so that he/she is free to teach during
the lesson and provide assistance as necessary in
making temporary mounts. The marking points below
are suggested for this assessment. If a pupil has
achieved all of these (or a major proportion if desired)
then he/she is awarded a "tick" for this skill in the
"master grid" for this objective.

- drawing made with sharp, clear lines
- drawing adequately large
- accurate representation of cells
- accurate labelling of all visible features
- neat labelling

Exercise 2.   Looking at Plant Cells

Introduction

If biologists want to look at plant cells they need to mount the cells on a microscope slide and examine them under the microscope. In this exercise you are going to make a temporary slide of some cells taken from an onion.

Making a temporary slide

The diagrams below and the instructions which go with them tell you how to make a temporary slide of some of your onion cells.

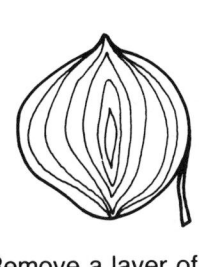

1. Remove a layer of flesh from an onion which has been cut open. Look for the very thin layer of cells which is now exposed.

2. Use a scapel or razor blade to cut a piece from this thin layer of cells. It should be about 0.5 cm × 0.5 cm big.

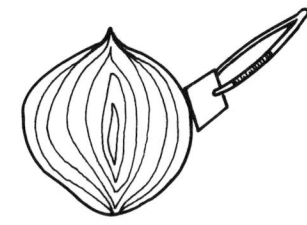

3. Remove the small piece of tissue you have cut with forceps.

4. Place the piece of tissue in the middle of a clean microscope slide.

5. Place a drop of iodine solution on to the tissue to stain it.

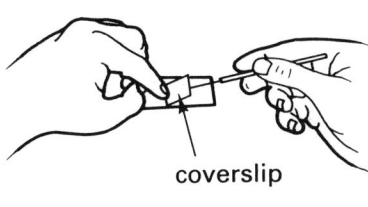

coverslip

6. Cover with a coverslip. Lower it carefully onto the slide. The stain will spread out.

Examining the cells

1 Put the slide onto the stage of the microscope.

2 Adjust the microscope so that light is shining through the slide and so that the cells are magnified enough to see inside them clearly. Adjust the focus knob so that the cells can be seen clearly.

3 Find a cell which is very clear and allows you to see all of the structures inside it easily.

Recording your observations

1 When you have found a good example of an onion cell, make a very clear drawing of exactly what it looks like.

2 Label your drawing as fully as you can.

Teacher's notes

Assessment objective

Ability to observe differences in fine detail
(NEA skills 9 and/or 10)

Previous experience

An understanding that organisms are put into groups of animals or plants with similar features would be an advantage.

Management strategy

Questions are presented requiring written answers

It is likely that this exercise would be taught in the context of a topic studying the diversity of living organisms. Ideally pupils would have studied the features of some taxonomic groups previously, and could then be expected to make a realistic attempt to find the features of the four Arthropod classes. The assessment of observation of differences is then based upon the written record of differences as pupils fill in their table.

Once pupils have filled in the tables, the next part of the activity is open to whatever interpretation the teacher wishes. Pupils could be given a set of pictures of different Arthropods to examine, but ideally a display of living and/or preserved specimens would be used.

Although intended as an exercise for assessment of the whole attainment range, it may be preferred for use only with average and above-average pupils since observation of differences in fine detail may be regarded as a high level skill.

Time required   Approximately ½ hour.

Resources required

Display of Arthropods: this may be a display of live or preserved specimens, a display of photographs of different Arthropods, a display of drawings of Arthropods, or a mixture of several of these.

Marking the assessment

The differences below are suggested as differences that pupils should be able to observe; these are differences in fine detail. However, the exercise could be used with low attaining pupils to assess the observation of differences in gross features.

If a pupil observes, say, two thirds of the *fine* differences, then the pupil is awarded a "tick" against the skill of observing fine details in the list of skills drawn up for the performance criteria of this objective. The differences shown in the list below are intended only as exemplars and should be modified as appropriate. Also, the proportion of the total number of differences which the pupils have to observe in order to have achieved competence in this skill can, of course, be varied to suit the requirements of the individual department.

It is important not to penalise pupils because they have not used the correct terminology to refer to features of the specimens. For example a pupil might call the abdomen of animal A (the dragonfly) a "tail" because that is what it looks like.

There is also the difficulty of deciding what is a gross feature and what is fine detail. It is suggested that a fine detail is an elaboration of the detail of a gross feature, eg the presence of eyes is a gross feature but the size of the eyes is a fine detail of the eyes.

- animal A has much larger eyes than animal B
- animal A has an abdomen with many segments; animal B has fewer segments
- animal B has much larger antennae than animal A
- the animals have different shaped "claws" at the end of their legs
- animal B has more segments in its legs
- animal B has two "teeth" sticking out from the front of its head but animal A does not
- animal A has a thorax which appears furry but animal B does not
- animal B's legs appear to be much more "furry" than the legs of animal A
- the first two segments of the front pair of legs of animal A are approximately the same length. The first segment of the legs of animal B is much shorter than the second segment

Exercise 3   Investigating Arthropods

Introduction

The group of animals called Arthropods have several pairs of jointed legs, and a hard external skeleton. Their bodies are made up of several distinct sections. There are four main groups of animals within the group called Arthropods. In this exercise you are going to try to find out what these groups are and what features an animal must have to be a member of one of these groups.

The diagrams below show two Arthropods that you may have seen before.

Animal A

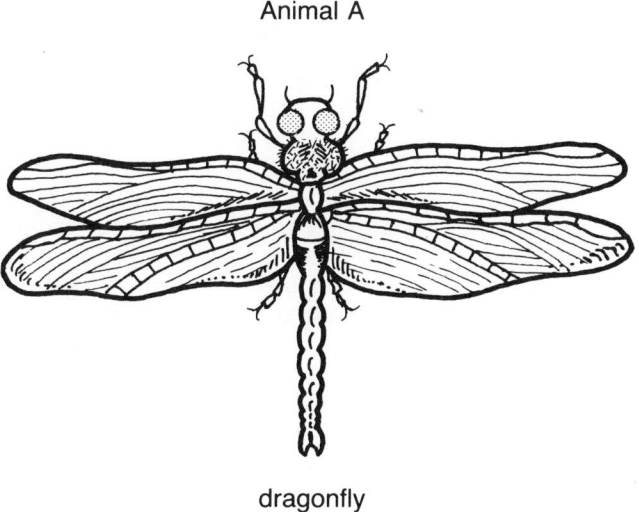

dragonfly

Animal B

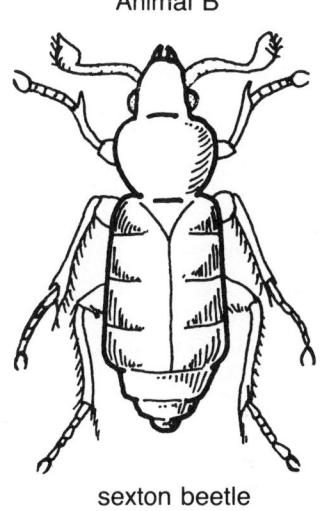

sexton beetle

What to do

1 Copy the table below and record any SIMILARITIES and DIFFERENCES between the two animals in the table.

Similarities	Differences

2 Study the pictures of the organisms carefully. The two animals belong to the same group of Arthropods. You have to try to decide why they are both in the same group. In other words, what features do they both have which put them in the same group?

3 Now you have found out what features the animals have in common, use whatever sources of information are available to you to decide which group of Arthropods these two animals belong to. Write down what you have found out.

4 Now look at the display of other animals belonging to the group Arthropods. Look up the features for each of the groups and try to put each of the Arthropods in the display into the right group. Record your findings in a table like the one below:

Animal	Group

Teacher's notes

Exercise 4

Assessment objective

Ability to design an experiment to test a hypothesis (NEA skills 29, 31 and 32)

Previous experience

Pupils must have a good understanding of the need of plants for light to carry out photosynthesis. The exercise could be used either in the context of a study of the stimuli that plants are likely to respond to, or at the end of a topic on photosynthesis. However, they must have no previous knowledge of tropisms.

Management strategy

This exercise is to be presented as a teacher demonstration.

It is suggested that the exercise begins with the pupils gathered around the teacher who begins by showing the class the plant which has grown towards the window of the dimly lit room in which it has been kept. The pupils should then be dispersed and arranged in seating positions as for a written test so that no collaboration is possible. The pupil sheets are then handed out and pupils can attempt the assessment.

It is intended that the exercise could be used across the whole attainment range. Differentiation is achieved by the stepped mark scheme since the pupils gain credit for what they include in their designs. It is clearly essential, if pupils are to achieve what they are really capable of, that the teacher makes very clear to the class the fact that assistance is available. This assistance should be in the form of standard clues, possibly written on "help" cards. Such clues might be:

Clue 1: Decide what apparatus you will need.
Clue 2: You need to set up two plants: one which is being tested to see if your hypothesis is correct and another as a comparison.
Clue 3: What factors must be kept constant to make the comparison between the two plants fair?
Clue 4: You need to keep everything which *might* affect the test plant constant, eg temperature, how much you water the plants, how much air movement there is around the plants, the soil the plants are growing in.
Clue 5: How long will you leave the plants?
Clue 6: What observations will you make and record?

Time required   ½–1 hour.

Resources required

A herbaceous plant which has been grown in unidirectional light so that it is showing a clear positive phototropism.

Marking the assessment

Since the exercise tests pupils' abilities to carry out all of the skills suggested in the performance criteria for the objective in Chapter 2, it is possible to award a grade for this objective from this exercise alone. Reference to the performance criteria shows that in order to obtain the highest grade, pupils should be able to:

(i) suggest several hypotheses which can be tested by experiment;
(ii) plan a suitable procedure to test a hypothesis;
(iii) take into account all of the variables to be controlled;
(iv) suggest adequate ways of controlling the variables;
(v) suggest measurements/observations that should be recorded.

The exercise expects pupils to suggest only one hypothesis. Therefore, although the whole attainment range could be presented with this exercise, it is intended to be used to assess up to a middle level of attainment.

Reference to the performance criteria will show how well each of these skills must be performed in order to achieve each of the grades attainable using this exercise. The criteria suggested in Chapter 2 show that even the lowest attainers can be assessed using this exercise because they have to be able to design a simple procedure to test a given hypothesis. Thus, if they are unable to suggest a hypothesis, the teacher can provide such pupils with one to test. Competence at each of the levels of achievement between these two extremes can similarly be accommodated within this exercise.

Teachers using the NEA scheme must decide at what point a pupil has received so much assistance that competence in skills 31 and 32 has not been demonstrated. Provided pupils were sufficiently prepared in these skills, especially skill 32, it would be justifiable to expect that all of the marking points are carried out without assistance if competence is to be recorded.

The marking points below are those suggested for this exercise. The assessor marks each pupil's design using these points and the pupils are awarded a "tick" for each skill for this objective in the master grid in which they show competence.

Marking Points

(a) Suggesting a plausible hypothesis in question 1

(b) suggesting appropriate procedure which will test the hypothesis, including:
   – suitable and available apparatus
   – appropriate time scale for experiment
   – suitable method of providing test conditions (eg bench lamp in a dark room to test if light is the factor making the plant bend)

(c) suggesting variables to be controlled:
   – temperature of environment around plant
   – water supply to plant
   – light regime of plant
   – air movements around plant

(d) suggesting methods of controlling variables

(e) suggesting observations which must be made

Exercise 4   The Bent Plant!

Your teacher will show you a plant which has been grown in a dark room. The plant has grown towards the window which has some light filtering through it and is bent as you see.

1 Write down what you think is the most likely hypothesis (idea) to explain what factor outside the plant you think has made it bend.

Now that you have a hypothesis to explain why the plant bent towards the window, the next stage is to design an experiment to test if the hypothesis is correct. You will be sat on your own to do this.

2 Design an experiment using the same type of plant shown to you to test the hypothesis that you wrote down in your answer to question 1.

IF YOU HAVE DIFFICULTY IN DESIGNING YOUR EXPERIMENT **AT ANY STAGE** ASK YOUR TEACHER FOR HELP SO THAT YOU CAN COMPLETE YOUR DESIGN.

Teacher's notes

Assessment objectives

Ability to measure: with a metric ruler
with a thermometer
with a stopclock

(NEA skill 1)

Previous experience

This exercise assumes that pupils are aware of the existence of pepsin and that finding the pH at which it works best is of value. Also required is an understanding that enzymes are responsible for digestion of biological chemicals and that protein is one such chemical. An understanding of the need for controls in experiments is needed, and it is assumed that pupils would be able to understand why the three control tubes are included in the experiment.

Management strategy

An additional assessor is required for this exercise.

It could be used as an inquiry exercise within the context of a study of the properties of enzymes, or of the conditions under which the enzyme pepsin in particular works. Since the suggested strategy will need planning and cooperation within the school, it is important that as many skills as possible within reason are assessed using the exercise.

Assessment of all or most of an average sized group would be possible during the exercise since the assessing teacher can devote all of his/her attention to the task. The teacher responsible for the assessment should ask pupils to show their ability to measure with a ruler whenever it is convenient before the pupils place the tubes into the water baths. (The pupil sheet also instructs pupils to ensure that the teacher has had the opportunity to assess their skill in measuring with a ruler *before* placing the tubes into the water bath.) Assessment of the skills of measuring with a stop-clock and with a thermometer can be carried out at any time during the exercise by asking pupils to measure the temperature of the water bath using the thermometer and the time on the stopclock. This will be especially easy if the stop clock can be stopped and restarted.

Neither of these measurements needs to be recorded for the purposes of the assessment.

Time required   Approximately 1 hour.

Resources required

Each of these items per pair of pupils:

6 test-tubes
Test-tube rack
Labels
Beaker for water bath
Tripod, gauze, bunsen burner and bench mat
Thermometer
Stopclock
1% pepsin solution
1 M hydrochloric acid
1 M sodium hydroxide
Distilled water
Egg white suspension (made by boiling egg white in a pan of water and then straining through muslin to produce a suspension of very small particles of boiled egg white)

Marking the assessment

The marking points below could be used in assessing the three skills within the objective of measurement. A "tick" is awarded for each skill in the master grid of skills for this objective drawn up for the performance criteria.

In the case of the NEA scheme this exercise could be used to assess two of the skills required for skill 1: measuring with a metric ruler and a thermometer.

Thermometer

Measured to an accuracy of within one degree on a scale marked in whole degrees.

Metric Ruler:

Measured to an accuracy of within one mm.

Stopclock:

Measured to within an accuracy of one division of the clock face (usually one second) or smallest division of time if a digital stopclock.

Exercise 5   What is the Best pH for Pepsin to Work?

Introduction

Pepsin is an enzyme which digests protein. It is found in the stomach, therefore it is likely that the pH in the stomach is the pH for pepsin to work best. This experiment is designed to test this hypothesis:

The pH in which pepsin works best is the same as the pH of the stomach.

The pH in the stomach is acidic. You are going to set up mixtures of protein and pepsin at different pH's to find out whether or not pepsin digests the protein most quickly in acid conditions.

Instructions

1 Set up a water bath at 37°C and keep it at this temperature throughout the experiment.

2 Place 6 test-tubes in a rack and number them 1 to 6 as shown below:

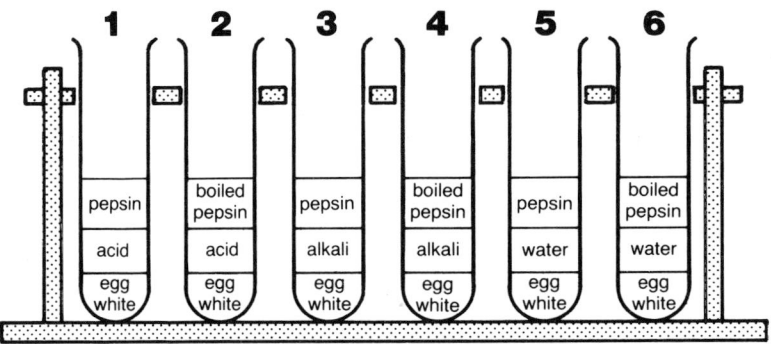

You are now going to measure out several solutions into the test-tubes as shown in the diagram above. While you are doing this your teacher will want to assess your ability to measure with a metric ruler.

3 Pour 1cm depth of egg white suspension into each of the test-tubes.

4 Pour 1cm depth of acid acid into two of the test-tubes; 1 cm depth of alkali into two of the test-tubes and 1cm depth of water into two of the test tubes.

5 Pour 1cm depth of pepsin solution into one of the tubes containing acid, one of the tubes containing alkali and one of the tubes containing water.

6 Pour 1cm depth of boiled pepsin solution into each of the three remaining test-tubes. These tubes are controls for the tubes containing pepsin to make sure that the protein is only digested when you put enzyme into it.

7 Shake the tubes well to mix the contents.

DO NOT GO ANY FURTHER UNTIL YOUR TEACHER HAS CHECKED YOUR APPARATUS

8 Place the six tubes in the water bath and start the stop-clock.

9 Record the time it takes for the egg white suspension to go clear, in each of the tubes in a table like the one below. Copy out this table while you are waiting for the pepsin in the tubes to go clear.

While you are doing this your teacher will ask you to measure the temperature of the water with the thermometer and also to measure the time on the stop-clock.

Tube number	Conditions	Time taken for suspension to clear
1	Acid	
2	Acid control	
3	Alkali	
4	Alkali control	
5	Neutral	
6	Neutral	

10 Examine the results in your table. In which pH conditions does it appear that pepsin works best? Is it the the same pH as in the stomach? If so, this means that the hypothesis must be correct.

Teacher's notes *Exercise 6*

Assessment objective

Ability to design an experiment to test a hypothesis (NEA skills 31 and 32)

Previous experience

Although no detailed theoretical knowledge is necessary for this exercise, it is likely that it would be taught within the context of work on enzymes, and a knowledge of enzymes might be needed to explain the results of the experiment. Other knowledge required is concerned with practical details such as how to maintain a solution at a particular temperature in a water bath.

Management strategy

Pupils will be carrying out activities during which they will be leaving a record.

Since the pupils are given the hypothesis to test, this exercise cannot be used to assess up to the highest level of achievement according to the performance criteria for this objective given in Chapter 2. The exercise is therefore intended for assessment of the lower end of the attainment range. In order to achieve differentiation the pupil sheet gives detail which guides pupils in drawing up their designs by making them first consider the answers to four important questions. The pupils would be seated as for a test to carry out the exercise. Again, pupils must be given assistance if necessary to answer questions (a) to (d). This must be taken into account in deciding if competence at a particular level has been achieved.

Whether or not the experiment is carried out is at the discretion of the teacher, but it is clearly desirable that it should be. The actual experiment could then be used to assess other objectives.

Time required ½ hour

Resources required

This equipment is required per pupil to carry out the experiment but not for the exercise itself. This list may need modification if pupils are to be permitted to carry out their own designs:

Piece of white cloth, approx. 8 cm x 4 cm (to be cut into two equal sized pieces 4 cm x 4 cm)
Samples of biological and non–biological washing powders
Beakers for water baths
Tripod, gauze and bench mat
Bunsen burner
Suitable stain (eg chocolate powder or gravy powder)
Thermometer

Marking the assessment

The criteria which must be satisfied to achieve the highest possible level of success in this assessment are given below. Most of the features required to satisfy the performance criteria for this skill can be assessed in this exercise.

Discrimination will be on the fullness of details, the range of variables which are going to be controlled and the depth of detail of how to control these variables. In short, this is an example of a stepped mark scheme since some of the marking points require a higher level of competence to include in the design than others. The marking points would be used as a checklist for each pupil's design, and then a "tick" awarded in the master grid for each skill achieved.

Pupils should include details of the following aspects of the experiment:

- type of material to be stained
- size of material to be stained
- nature of stain
- how much stain will be used
- how stain will be applied
- statement that stains on pieces of material to be compared must be same size and contain the same amount of staining substance
- amount of each powder that will be dissolved in the washing water
- volume of washing water to be used
- container in which washing will take place
- temperature of wash
- statement that temperatures must be the same for both powders
- washing procedure
- rinsing procedure
- statement that washing procedures and rinsing procedures must be the same for each powder
- method of comparing how clean cloths are after washing
- observations which will be recorded

Exercise 6   Does a Biological Washing Powder Wash Whiter at Low Temperatures?

Introduction

You will have seen advertisements claiming that so-called "biological" washing powders can wash whiter at lower temperatures. We can make a hypothesis that could be tested by experiment:

A biological washing powder will wash cleaner than a non-biological powder at low temperatures.

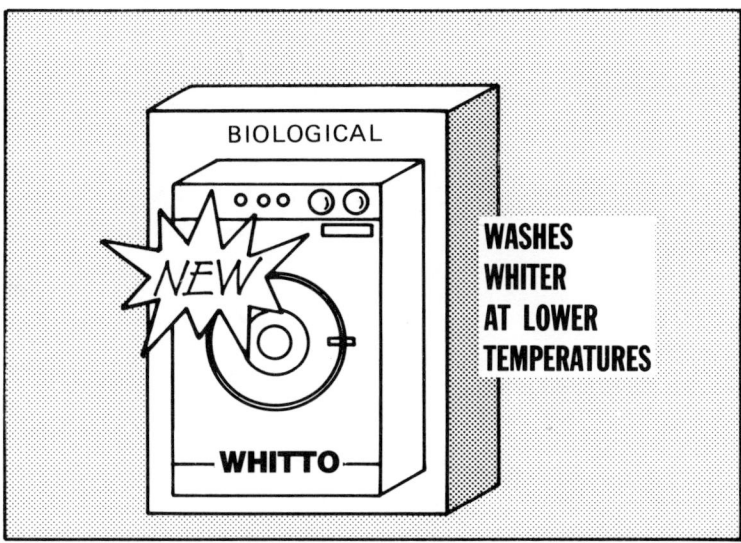

YOU WILL BE SAT APART FROM OTHER PUPILS TO DO THIS EXERCISE.

You have to decide on the answers to several questions before you can test this hypothesis:

(a) Which powders are biological and which are not?

(b) What does "cleaner" mean, and how could you measure whether something is cleaner than something else?

(c) What should you try to wash to test this hypothesis?

(d) What are "low" temperatures?

Instructions

1. Decide what you think the answers to these questions should be. If necessary get help from your teacher to answer them. Write down your answers.

When you have sensible answers to these questions you have the information you need to design an experiment to test the hypothesis.

2. Design an experiment you could do to test the hypothesis. Write down your design.

MAKE SURE that you describe the design in detail which is good enough for somebody else in your group to do the experiment.

Teacher's notes

Assessment objective

Ability to follow instructions
(NEA skill 12)

Previous experience

No previous experience is essential to carry out this exercise. It is likely, however, that the exercise would be used in the context of work on the balanced diet and that pupils will understand the need to examine foods for their nutrient content.

Management strategy

The exercise requires the setting up of four stations.

It is intended that each pair of pupils tests a different food for each of the nutrients and that the results of the class be brought together at the end. It will be necessary for pupils to draw up their own results table before beginning. Pupils visit each of the stations and test their food at each one for each nutrient. They are then instructed to clean up their station before moving on to the next.

The performance criteria suggested for this objective in Chapter 2 include the assessment of the ability to follow branched instructions. None of the tests include such instructions, therefore this exercise would be used to assess at middle and lower levels of attainment. For this reason the instructions have been given in highly diagrammatic form. Nevertheless, the exercise could be used to assess higher attaining pupils by modifying the tests so that, for instance, they have to find out the most easily digested form of carbohydrate in each food. If they do this then they could start with the sugar test: the instruction would say that if the test for sugar proved positive then they would go on to a protein or fat test but if there was no sugar they would test for starch, the next carbohydrate.

It may be possible using this exercise to assess the whole class on their ability to follow instructions. In order to establish fairness, it would be desirable to ensure that all pupils are assessed on their ability to follow the instructions to carry out *one* of the tests, say, the sugar test. Teachers may feel, however, that some of the tests are of equal difficulty and that any of these could be used to assess the objective. If only a proportion of the class can be assessed using the exercise, then their achievements would be recorded in the master grid for this objective and the others assessed at a later date.

Time required Approx. 1 hour.

Resources required

Station 1: Starch Test
Iodine in potassium iodide solution
Spotting tile
Pestle and mortar
Teat pipette

Station 2: Sugar Test
Benedict's solution
Pestle and mortar
Boiling tubes
Test-tube rack
Test-tube holders
Goggles
Bunsen burner
Bench mat

Station 3: Protein Test
Biuret reagents (sodium hydroxide and copper sulphate)
Pestle and mortar
Test-tubes
Test-tube rack

Station 4: Fat Test
70% ethanol
Pestle and mortar
Test-tubes
Test-tube rack
Distilled water

Each station will also need a waste food container.

Ideally there should be 3 or 4 of each of these stations available to pupils so that each pair of pupils can always gain access to the equipment to carry out a test. This may require having one or two more stations than there are pairs of pupils.

Marking the assessment

The marking of the assessment is straightforward but requires the teacher to observe the pupils carrying out the tests. Pupils will probably be working in pairs, but for assessment of an individual it is important that the pupil does the test being assessed without the partner. A pupil who is able to follow a sequence of instructions without assistance will have achieved competence at the middle level of attainment according to the performance criteria given in Chapter 2. If a pupil requires each instruction to be given one at a time, the teacher is on hand to provide the assistance required, and the pupil is successful at the lowest level of attainment. Thus each pupil is measured against the performance criteria which have been drawn up by the teacher or department for this objective and awarded the appropriate grade.

Exercise 7   Investigating the Nutrient Content of Foods

Introduction

Because it is important that the diet we eat is a balanced one, it is important to know which foods contain which nutrients so that we can balance our own diet. In this exercise you are going to test several foods to find out which nutrients they contain.

The nutrients you are going to test for are: **Sugar   Starch   Protein   Fat**

Instructions

The foods that you are going to test have been divided up between the members of the class. You and your partner have one food and you are going to test it for all of the nutrients.

The tests will be carried out at the stations around the room. You have to take samples of your food to the stations to do the test.

The instructions below show you how to do each of the tests on your food. Follow the instructions shown in the diagrams. Clean apparatus thoroughly between tests.

WHILE YOU ARE DOING THIS YOUR TEACHER WILL CHECK HOW WELL YOU ARE FOLLOWING THE INSTRUCTIONS.

Testing for Sugar

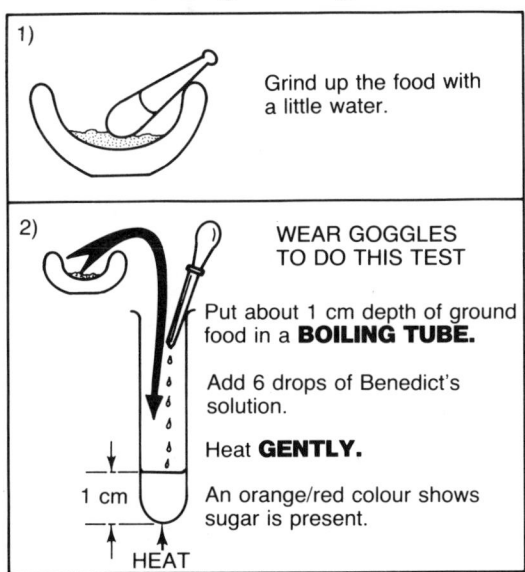

1) Grind up the food with a little water.

2) WEAR GOGGLES TO DO THIS TEST

Put about 1 cm depth of ground food in a **BOILING TUBE.**

Add 6 drops of Benedict's solution.

Heat **GENTLY.**

An orange/red colour shows sugar is present.

1 cm

HEAT

Testing for Starch

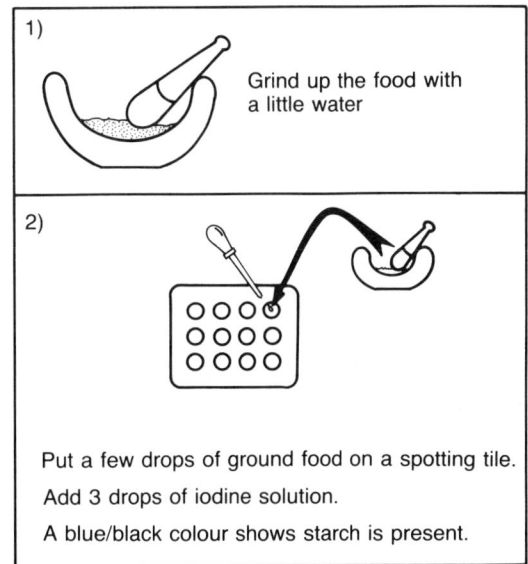

1) Grind up the food with a little water

2) Put a few drops of ground food on a spotting tile.

Add 3 drops of iodine solution.

A blue/black colour shows starch is present.

Testing for Protein

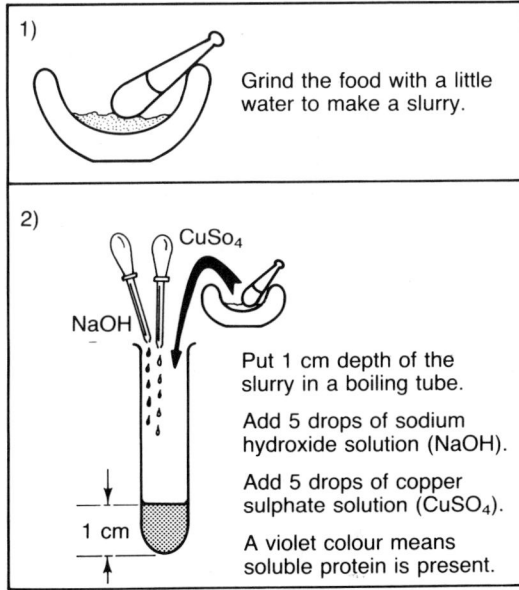

1) Grind the food with a little water to make a slurry.

2) CuSo₄

NaOH

Put 1 cm depth of the slurry in a boiling tube.

Add 5 drops of sodium hydroxide solution (NaOH).

Add 5 drops of copper sulphate solution ($CuSO_4$).

A violet colour means soluble protein is present.

1 cm

Testing for Fat

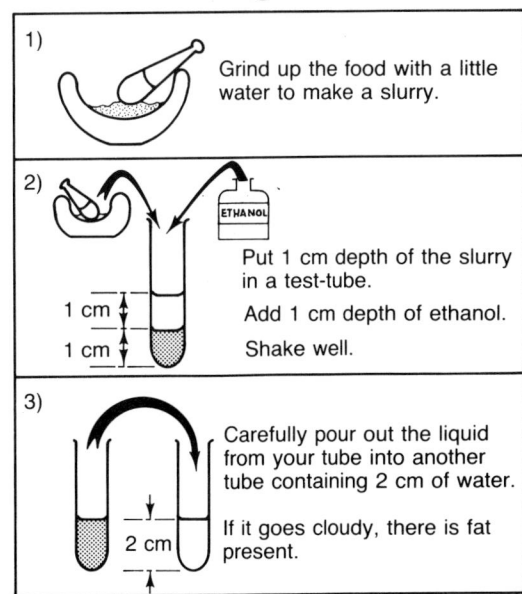

1) Grind up the food with a little water to make a slurry.

2) ETHANOL

Put 1 cm depth of the slurry in a test-tube.

Add 1 cm depth of ethanol.

Shake well.

1 cm

1 cm

3) Carefully pour out the liquid from your tube into another tube containing 2 cm of water.

If it goes cloudy, there is fat present.

2 cm

Teacher's notes

Assessment objective

Ability to interpret results
(NEA skills 20, 21, 22, 25 and 26)

Previous experience

Pupils will require knowledge of the process of respiration and of the effect of temperature on enzyme activity. Thus this exercise might be used as an activity to draw attention to an industrial application of the knowledge gained at the end of a topic on respiration in which the importance and properties of enzymes has been stressed.

Management strategy

It is not intended that pupils necessarily carry out this experiment since the exercise is an illustration of an assessment using questions requiring written answers. A simple experiment has been chosen, however, so that pupils could carry out the experiment if this is desired, and the ability to interpret results could be assessed using pupils' own results. It is suggested that pupils are seated as for a test.

The exercise is intended for assessment of pupils of above average attainment but simplification of the tasks and results table could make it suitable for use with lower attaining pupils.

Time required   Approx. 1 hour

Resource required

Graph paper.

Marking the assessment

The questions on the sheet have been chosen to ensure the assessment of all the skills suggested in the performance criteria for this objective in Chapter 2. Each of the questions assesses a different skill within the objective of interpreting results. The exercise is broadly stepped so that more difficult questions are asked after simpler tasks have been carried out. Therefore all that is required is for a teacher to mark pupils' answers to questions. If the question is correctly answered then a "tick" can be placed against the pupil's name in the master grid for that skill. After all answers have been marked for a pupil, the teacher awards a grade for the objective in accordance with the performance criteria.

It is important that, where necessary, pupils are given the axes for the graph they are expected to draw. Correct answers to any questions they feel unable to tackle should be given to achieve differentiation. These could be provided on "help" cards or sheets. The axes for the graph and the answers to each of the questions could be put onto separate pieces of card or sheets, and made available to pupils if they need them. The teacher then records which pupils took advantage of this help since such pupils will not have demonstrated competence.

The answers to questions are listed below, together with the skill which the question is intended to assess. It is a matter of professional discretion to decide how near to the detail given the pupils must get to achieve competence. Since the ability to draw a graph is not assessed in this exercise, pupils must not be penalised because they have different answers from those in the list below if they result from poor graph drawing, provided that the answers given are correct for the graph they have drawn.

Q1 40°C.
(Ability to read results from a table)

Q2 26.3.
(Ability to perform simple calculations)

Q3 Number of bubbles increases at first, then decreases.
(Ability to identify patterns)

Q4 An increase in temperature causes an increase in the number of bubbles produced by a yeast up to 35°C–40°C.
(Ability to draw conclusions)

Q5 An increase in temperature causes an increase in the rate of enzyme reactions in the yeast. This increases the rate of respiration which increases the rate of production of bubbles of carbon dioxide.
(Ability to explain conclusions in terms of fundamental principles)

Q6 27–28.
(Ability to interpolate from graphs)

Q7 Zero.
(Ability to extrapolate from graphs)

Q8 Total before 40°C = 92
Total after 40°C = 128
Therefore percentage = $(92/128) \times 100 = 71.9\%$.
(Ability to perform simple calculations including percentages)

Exercise 8   Reducing Sugar Content in Waste Water

The problem

Many sweet manufacturers have a problem with waste water when they are making sweets. The water used in the process of making the sweets often has a high sugar content. If they allow this water to run away, then it may pollute the streams or rivers into which it flows. They must therefore attempt to reduce the sugar content before they can allow it to flow away. How can they do this?

One answer to the problem is to use the sugar to grow yeast. The yeast cells multiply as they use up the sugar in the water, and the large amount of yeast produced can be sold to farmers as animal feed.

This still leaves the sweet manufacturers with the problem of finding a way of setting up the yeast and waste water so that the yeast can feed off the sugar. Also what are the best conditions to use to set up the process: What temperature? What pH? What sugar concentration?

Your task

Your task is to try to find the answer to the first of these questions: What is the best temperature to set up the process of growing yeast on the sugar in waste water?

The apparatus shown here could be used to carry out the experiment. As the yeast uses the sugar, it produces carbon dioxide gas which bubbles up through the water. The number of bubbles being produced in a given time could therefore be counted.

If the tube containing the yeast and sugar solution is placed in a water bath at a constant temperature it would be possible to find out how many bubbles of carbon dioxide gas are given off at a certain temperature. The temperature of the water bath could then be changed and the experiment repeated.

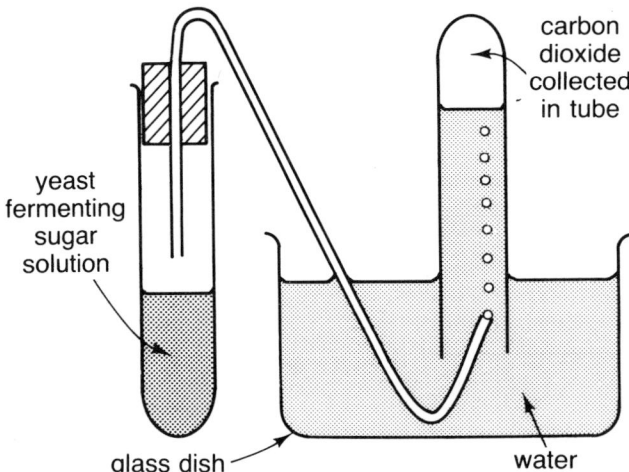

yeast fermenting sugar solution

carbon dioxide collected in tube

glass dish

water

Interpreting the results

Not surprisingly, this experiment has already been done by sweet manufacturers. The results of three experiments are shown in the table below.

Plot three line graphs for these results on one piece of graph paper.

ASK YOUR TEACHER FOR HELP IN DRAWING THE AXES FOR THE GRAPH IF NECESSARY.

Temperature	Number of bubbles produced per minute		
	Experiment 1	Experiment 2	Experiment 3
10	5	4	6
15	7	8	9
20	11	14	16
25	16	20	24
30	23	26	30
35	30	34	36
40	32	35	36
45	4	4	2

NOW COMPLETE THESE TASKS. If you have any difficulty with any of them ask your teacher to help you.

1 Study the table. At what temperature are most bubbles produced in Experiment 1?

2 What is the average number of bubbles produced during the three experiments at a temperature of 30°C? Show your working.

3 Study your graph. What pattern do you see in the number of bubbles produced per minute as the temperature is increased?

4 Study your graph. Describe what conclusion you draw from your results about the effect of temperature on the rate of bubble production by yeast.

5 Explain why an increase in temperature from 15°C to 30°C has the effect that you have described in question 4.

6 What would be the number of bubbles produced by the yeast at 28°C in Experiment 1?

7 What would be the number of bubbles produced by the yeast in Experiment 2 at 50°C?

8 How many bubbles were produced in all of the tests up to (but not including) 40°C in Experiment 1? Work out what percentage this is of the total number of bubbles produced in *all* of the temperatures for experiment.

Teacher's notes

Exercise 9

Assessment objective

Ability to measure: with a metric ruler
                with a measuring cylinder
                with a stopclock
(NEA skills 1 and 3)

Previous experience

Knowledge of the process of diffusion is important. An understanding of pH indicators, acids and alkalis, whilst not essential, would be very desirable since the concept load of this exercise is likely to be too high for many pupils unless they have sufficient previous knowledge to understand that the jelly has been dyed red by a pH indicator and will turn yellow because that is the colour that the indicator turns in acid conditions.

Management strategy

Within the context of routine coursework on gaseous exchange, diffusion or transport, this exercise could serve as an enquiry activity which could be used at different levels. For higher attaining pupils it could be used as an introduction to a consideration of the effect of surface-area-to-volume ratio on the effectiveness of diffusion. For lower attainers the concept level of the exercise itself could be the correct level to approach the subject. Thus the activity could be used across the whole attainment range.

Since the management strategy is to assess using normal practical coursework, the teacher would not necessarily aim to assess all pupils. For this reason the pupil sheet does not instruct pupils to ask the teacher to go over to them to watch as they are measuring. The teacher would want to assess those pupils that he/she observes measuring. The pupil sheet alerts pupils to the fact that they MAY be assessed. Those pupils who are assessed would therefore gain a "tick" against their name in the master grid for the three skills assessed here.

Time required   Approx. 1 hour

Resources required

Per pupil or pair of pupils:
Boiling tube and bung
2 cm × 2 cm × 2 cm block of gelatine jelly dyed red with cresol red indicator and ammonium hydroxide
1 M hydrochloric acid
Scalpel or knife
Measuring cylinder (calibrated in 1cm^3 units)
Stopclock

Marking the assessment

The marking points below could be used in assessing each of the three measuring skills assessed using this exercise.

Metric ruler
Measured to an accuracy of within one mm.

Measuring cylinder
Measured to an accuracy of within one cm^3 at the bottom of the meniscus.

Stop-clock
Measured to within an accuracy of one division of the clock face (usually one second) or the smallest division of time on a digital stopclock.

Exercise 9   Diffusion and Size

Introduction
Here are some organisms which are all different sizes in real life:

All of them are made of cells and all of the cells need to have some substances to stay alive. Substances like oxygen and water have to get to all of the cells. One way that this could happen is by diffusion. This raises a question, however: "Is diffusion quick enough to allow substances to get to all of the cells in an organism?"

In this exercise you are going to try to answer this question. Instead of organisms of different sizes you are going to use blocks of jelly of different sizes. This is because the jelly can be stained red and when an acid diffuses into it the acid turns the block yellow so that you can see where it has got to.

Instructions
1 Cut a block of red jelly into 5 cubes of different sizes. The actual sizes do not matter, but you must measure the size of each cube and record the size of each cube in the results table.

WHILE YOU ARE MEASURING THE CUBES OF JELLY YOUR TEACHER MAY COME ROUND TO ASSESS YOUR ABILITY TO MEASURE WITH A RULER.

2 Copy down the results table shown   below.

Length of sides of cube (mm)	Time taken to turn yellow (s)

3 Pour 20 cm^3 of acid in a boiling tube.

WHILE YOU ARE MEASURING OUT THE ACID YOUR TEACHER MAY COME ROUND TO ASSESS YOUR ABILITY TO MEASURE OUT WITH A MEASURING CYLINDER.

4 Place a cube of each size in the boiling tube. Start your stopclock.

5 Put the bung in the top of the tube.

6 Lay the tube on its side and try to spread out the cubes of jelly as shown:

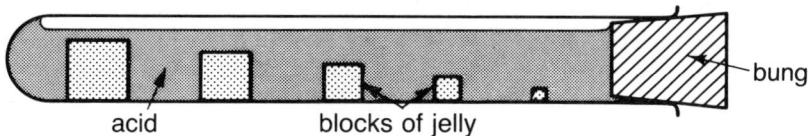

acid   blocks of jelly   bung

7 Use the stopclock to time how long it takes for each cube to turn completely yellow. Record the time when each cube turns yellow in the results table.

WHILE YOU ARE MEASURING THE TIME IT TAKES FOR THE ACID TO DIFFUSE INTO THE JELLY, YOUR TEACHER MAY COME ROUND TO ASSESS YOUR ABILITY TO MEASURE WITH A STOPCLOCK.

8 Examine your results carefully and decide if you can see a pattern in them between the size of the cube and the time it takes for the acid to diffuse into the whole cube.

Teacher's notes *Exercise 10*

Assessment objective

Ability to follow instructions.

Previous experience

No previous experience is essential, although it would be desirable for pupils to be aware of the importance of plants as producers in ecosystems in order that they appreciate the significance of the exercise.

Management strategy

An additional assessor is required in this exercise. The role of the additional assessor would be to supervise during the potentially hazardous burning of the plant material. An additional role of the additional teacher is to collate class results. The class teacher is thus freed to assess pupils by observing them following instructions.

It is suggested that the whole exercise is carried out as a class activity in which the results of each pupil are collated at the end to provide a set of results which could be used to find the mean energy content of the plant material under investigation. Alternatively, and perhaps more usefully, different plant communities could be compared by providing different pupils with plant material.

The exercise can be carried out safely provided a class is well prepared and that clear instructions in the event of accidents are given at the start of the lesson. Notwithstanding this precaution, there *is* inherent danger in burning plant material and the exercise should not be carried out if the teacher does not feel confident about the expected behaviour of pupils.

There is enormous variability between plant samples and between determinations of the energy content of plant samples using this technique, and it may therefore be considered important to stress this point. In this case pupils could be instructed to carry out three determinations of plant material of a single type by following the instructions three times. All three readings would then be recorded in the class results. Time may not be available for this extra teaching point, however, therefore the pupil sheet does not confuse pupils by giving them an instruction that they may not carry out.

The exercise provides the opportunity for pupils to show competence in all skills associated with this objective except the ability to follow branched instructions. The exercise is intended to be used to assess pupils of up to the highest level below that involving branched instructions since the calculations involved may not enable lower attaining pupils to achieve their true level of success in this activity.

It is important that assistance is provided in carrying out any of the stages of the method, such as assembling the clamp stand, boss and clamp, since inability to perform manipulative tasks, calculations, measurements, etc. must not prejudice a pupil's chance of success in the objective being assessed.

The additional teacher would have a role of both manager of the class, and collator of class results in this exercise, while the assessing teacher would attempt to assess as many pupils as possible.

Time required   Approx. 1 hour.

Resources required

Per pupil:

Clamp stand, clamp and boss
Tin can
Measuring cylinder
Metal dish, tin lid or other suitable container for burning plant material
Bench mat
Thermometer
Any suitable dried plant material, eg chopped grass
Access to a balance.
Goggles

Marking the assessment

The following list represents the sequence of operations which pupils should perform to be awarded the highest level of attainment in this exercise in accordance with the performance criteria suggested for this skill in Chapter 2. Lower grades would be awarded if the pupil can follow a proportion of the instructions unaided, and the lowest level of attainment would be awarded to pupils able to carry out one instruction.

- apparatus set up first
- 100 cm^3 of water placed in tin can
- dried plant material placed in metal container
- plant material burned with can as close as possible to plant material
- energy content of plant material per g given to teacher.

Recordings made

- weight of plant material recorded
- initial temperature of water
- final temperature of water
- temperature difference
- energy content of plant material per g (Calculation must not be assessed, therefore the accuracy of this value is unimportant)

Exercise 10 How Much Energy is Contained in a Plant Community?

Introduction

The organisms living in almost any area rely entirely on the plants in the area to provide them with the energy they need to live. Carnivores, for instance, obtain their energy by eating herbivores and herbivores obtain their energy by eating plants. The number of organisms that can live in an area depends upon how much energy is available to them from the plant community. In this exercise you will use a simple method to find out how much energy is contained in a plant community.

Instructions

READ **ALL** OF THESE INSTRUCTIONS BEFORE STARTING

These instructions tell you how to set up your apparatus and carry out the experiment. When you have finished, your teacher will assess your ability to follow instructions by looking at what you have done.

MAKE SURE THAT ALL THE RECORDINGS YOU MAKE ARE CLEAR FOR THE TEACHER TO SEE.

WEAR GOGGLES DURING THIS EXERCISE.

1 Set up the apparatus as shown in the diagram opposite.

ASK YOUR TEACHER FOR ANY HELP YOU MAY NEED TO DO THIS.

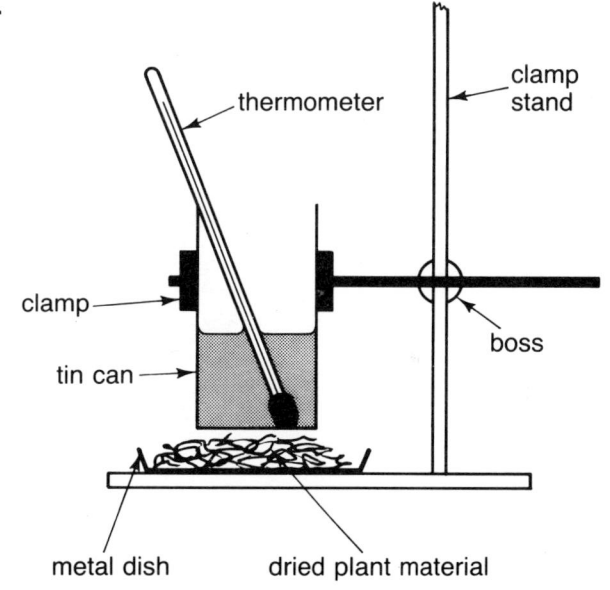

2 Put 100cm^3 of water into the can.

3 Weigh out some dried plant material and place it in the metal dish.

4 Write down the mass of plants that you have placed in the dish.

5 Write down the temperature of the water in the can.

6 Light the plant material and let it heat the water in the can. Make sure that the can is as close as possible to the plant material without putting the burning plant material out.

7 When the plant material has finished burning, record the temperature of the water.

8 Take away the first temperature from the final temperature and record this temperature difference.

9 A simple calculation can be used to find out how much energy was contained in the plant material:

Amount of energy in 1g of plant material $= \dfrac{420 \times x}{m}$

where $x =$ temperature rise of water in your experiment,
    $m =$ mass of plant material you burned.

10 Give this figure to your teacher to record with the class results.

Teacher's notes

Exercise 11

Assessment objective Ability to record results (NEA skills 15, 16 and 17)

Previous experience

No previous experience is essential; this exercise could be used as an introduction to a topic on senses and/or receptors in humans.

Management strategy

This is a stations exercise. With the instructions on the sheet and the equipment available at each of five stations, the activity should be self-running. The investigation would need to be set up, and at the outset the line in the introduction stating that the pupils will have to think about what they have learned at the end will need to be stressed, to ensure that pupils are thinking as they carry out the activities. It is likely that time will be available at the end of the activity to draw out these points for each activity:

Taste: Different areas of the tongue are sensitive to different tastes and only one taste per area.

Sight: There is a limit to how well we are able to make out details using unaided vision. Some people are better than others at identifying details at distance. (The term "resolution" could be used here.)

Hearing: There is a limit to the range of frequencies that the human ear can detect. Different people have different ranges.

Touch: Some parts of the body are able to detect two points when they are closer together than other parts can.

Hot and cold: We can detect the difference in temperature between two liquids/objects but we cannot say what the actual temperature is using our skin. (It may be, of course, that a few pupils will estimate the temperature of the water very accurately. Nevertheless these will be exceptions and the general point about temperature sensations can still be made.)

Since the abilities to fill in a table and to draw a suitable table and fill it in are both assessed using this exercise it is important that the teacher draws pupils' attention to this fact before they start and then provides such assistance as may be necessary to ensure that all pupils can draw a table. Duplicated copies of the tables could be available as "help" cards or sheets for pupils who require them. Provided differentiation is ensured in this way it is intended that this activity be used across the whole attainment range.

Some activities must be carried out in pairs. In this case the experimenter can be the recorder. An important function of the teacher must therefore be to prevent pupils helping each other with their recording.

Time required   Approx. 1 hour

Resources required

There are five different activities, each one requiring the equipment below for one station. It is desirable to make at least two of each station available so that all pupils are engaged in useful activity at all times and to keep the time required to the minimum.

Taste
Four small beakers, each labelled (eg A to D):
– beaker containing bitter solution, eg coffee or hops extract
– beaker containing sour solution, eg lemon juice
– beaker containing sweet solution, eg sugar
– beaker containing salt solution
5cm × 1cm strips of filter paper
Waste paper container
Distilled water (preferably in a bottle)
Waste water container (or station near a sink)

Sight
Set of cards of any size on which are a large number of the following letters of standard size. A stencil is the easiest way of ensuring that the letters are all of the same size. Suggested letters are those below since they are very similar: **G O C Q M N W H**

Hearing
Frequency generator capable of producing frequencies which exceed both the highest and lowest frequencies audible to the human ear
Loudspeaker (or headphones if available)

Touch	Hot and cold
Straws	Beakers containing hot, cold and warm
Scissors	water (these will need to be replen-
Waste paper	ished at intervals during the lesson)
container	

Marking the assessment

It is intended that this exercise be used to assess three of the recording skills drawn up for the performance criteria for the objective of recording results. Viz:

Recording results in the form of diagrams (Taste station)
Recording results in prepared tables (Touch station)
Drawing up a table and recording results in it (Hearing and/or Hot and Cold stations)

The marking points below are suggested for each of these skills. Pupils are marked using these points and a "tick" placed against their name in the master grid for this objective for any of the skills which they achieve. Teachers will need to decide how many of these marking points, if not all of them, must be satisfied if a pupil is to be deemed to have shown competence in the skill.

Recording results in the form of diagrams:
– drawing of tongue map with clear, sharp pencil lines
– drawing accurately copied
– large drawing
– areas of map labelled neatly

Recording results in prepared tables:
– information recorded in correct columns
– units correct
– recordings made neatly

Drawing up a table and recording results in it:
– appropriate columns
– table drawn neatly
– columns headings correct
– units correct
– recordings made neatly
– recordings made in correct columns

Exercise 11   Investigating Human Senses

Introduction

Around the room there are a number of stations at which you have to carry out short activities. The instructions for each activity are given below. The activities will give you some information about the following human senses:

Taste   Sight   Hearing   Touch   Heat and cold sensation

When you have finished you will need to think about what you have done and write down what you have learned.

Instructions for stations

RECORD YOUR OBSERVATIONS AS YOU GO ALONG

Taste

1 Copy the diagram shown opposite. This is a map of your tongue showing which areas of taste the tongue is divided into.

2 Find out which areas are sensitive to which tastes using these instructions. Record your observations neatly on your map of the tongue.

3 Take a strip of filter paper and dip it into one of the five solutions at this station. Each one is labelled to identify it.

4 Place the wet end of the paper onto each of the areas of your tongue shown on the map.

5 Record on the map which areas are sensitive to this taste.

6 Wash out your mouth thoroughly with distilled water.

7 Repeat the test for each of the other tastes.

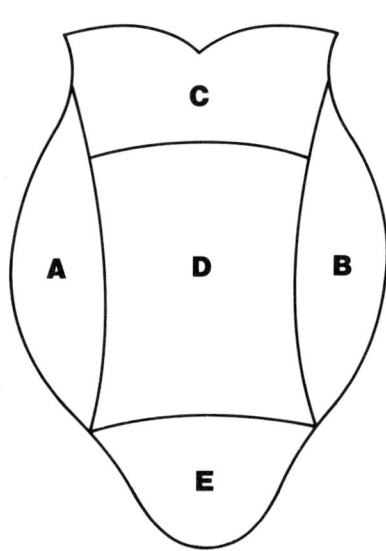

Sight

COPY THE RESULTS TABLE TO RECORD YOUR RESULTS IN:

Letter	Furthest distance seen

1 This station has a set of cards with letters that look very similar.

2 Your partner should pick each card in turn without you knowing which has been picked. Your partner should then bring each of the cards towards you from the opposite side of the room. Record in the table the distance that each card is from you when you can just make out what the letters are.

Hearing

YOU WILL HAVE TO DRAW OUT A TABLE TO RECORD THE RESULTS OF THIS ACTIVITY. READ THROUGH THE INSTRUCTIONS AND THEN DRAW THE TABLE BEFORE YOU BEGIN.

1 This station has an instrument called a frequency generator linked to a loudspeaker. This instrument can make sounds at different pitches.

2 Turn the pitch control on the frequency generator to the highest setting. Turn the control so that the pitch gets lower and lower. Note the setting on the control when you can first hear the sound and record it in your table.

3 Keep turning the control until the pitch is so low that you cannot hear it any more. Record the setting on the control where this happens in your table.

Touch

COPY THE RESULTS TABLE BELOW TO RECORD YOUR RESULTS IN.

Area of body	Smallest gap between points

1 Cut across the ends of a straw so that you have two points as shown below:

2 Bend the straw in half so that the two points can be brought together.

3 Survey the skin in different parts of the body to find the smallest gap you can have between the points and still tell that there are two separate points. Parts of the body you could investigate are: the back of the hand; palm of the hand, fingertips, leg, neck, cheek etc.

4 Record your measurements in the results table.

Hot and cold

YOU WILL HAVE TO DRAW OUT A TABLE FOR THE RESULTS OF THIS ACTIVITY. READ THROUGH THE INSTRUCTIONS FIRST AND THEN DRAW OUT YOUR TABLE.

1 This station has beakers of water at different temperatures. One is hot, one is warm and one is cold.

2 Use your fingers to decide which is which. Record your findings in the table.

3 Now try to estimate the actual temperature of the water in each of the beakers. Record these temperatures in the table also.

Teacher's notes *Exercise 12*

Assessment objective

Ability to interpret results
(NEA skills 20, 21, 25, 26 and 29)

Previous experience

It is important that pupils understand why the apparatus can be used as a model for a human body being cooled with and without sweating. This concept level is likely to require explanations of different lengths for different groups of pupils, and for this reason an explanation of how the apparatus models humans has not been included on the pupil sheet.

It is likely that this exercise would be used within the context of a topic investigating thermoregulation in mammals and therefore the knowledge necessary is that humans do sweat in order to cool down.

Management strategy

This is an exercise which leaves a record. The pupils' sheet is self explanatory. It is important, however, that pupils are given any assistance required in carrying out the experiment, recording results and plotting their graphs to avoid prejudicing their achievement in interpreting their results. The exercise has been chosen to assess this objective because pupils can gain useful information about the pattern of results from a graph of the results that they would find extremely difficult to gain from the table of results alone.

It is suggested that pupils work alone to avoid collusion. The consequent difficulties in supply of equipment that may arise are balanced by the fact that the exercise allows the opportunity for assessment of as many of the skills drawn up for the performance criteria for this objective as possible. Further, the choice of management strategy ensures that the whole class can be assessed.

To achieve differentiation, the questions are stepped within the limits set by having a logical sequence of questions. It is intended that this exercise could be used across the whole attainment range provided adequate assistance is provided to pupils in achieving the skills under assessment. If this is not possible, the exercise should be used only with higher attaining pupils or the questions intended to test their abilities in the skills for this objective should be simplified by removing those testing the highest level skills.

Time required   Approx. 1 hour

Resources required

Per pupil:
Two test-tubes or boiling tubes of equal size
Thermometer
Stopclock
Newspaper
Elastic bands or string
2 clamp stands, clamps and bosses
Supply of hot water

Marking the assessment

The questions on the sheet have been designed to assess most of the skills suggested in the performance criteria in Chapter 2 for the objective "ability to interpret results". The only skills not assessed here are those concerning performance of calculations.

The marking points given below would be used to decide each pupil's level of competence in each skill. Each question tests a different skill, therefore for any answers which are correct a pupil will gain a "tick" for that skill in the master grid for this objective. It is important that pupils are marked in accordance with the mark scheme suggested and that their answers must comply fully with the scheme before they are deemed correct (for example, missing out units would prevent pupils achieving success). By marking pupils' work after the lesson, the whole class can be assessed on each of these skills.

Q7 Answer accurate with correct units.
(Ability to read corresponding values from tables)

Q8 Answer correct +/– one division of graph paper with correct units.
(Ability to interpolate values from graphs)

Q9 Water in wet tube cools down.
(Ability to identify patterns in data)

Q10 Water in dry tube cools down more slowly than water in wet tube.
(Ability to identify patterns in data)

Q11 A hot object with a wet surface will cool down more quickly than a hot object with a dry surface.
(Ability to draw conclusions from results)

Q12 The wet surface of the wet tube cools down the water inside the tube more quickly than does the dry surface because it takes heat energy to evaporate the water from the surface.
(Ability to explain conclusions in terms of underlying theory).

Q13 Answer correct +/– one division of the graph paper with correct units.
(Ability to extrapolate information from a graph)

Exercise 12 How Effective is Sweating at Cooling you Down?

Introduction

You know that when you get hot you sweat. This is your body's way of trying to cool you down. But how effective is sweating in doing this? In this exercise you are going to try to find the answer to this question.

Instructions

1 Set up the apparatus as shown in the diagram below.

MAKE SURE THAT THE PAPER AROUND THE TUBES IS THE SAME THICKNESS ON EACH TUBE

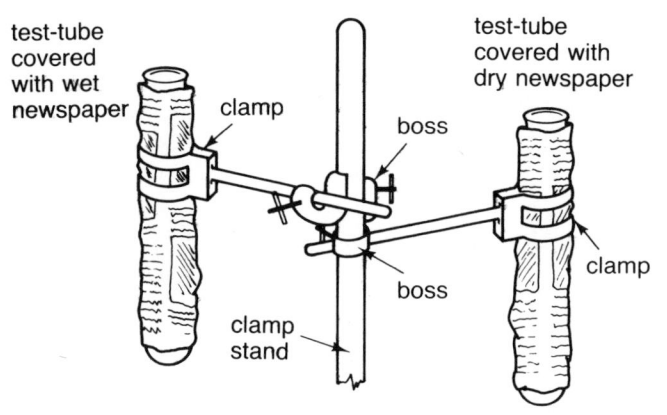

2 Copy the results table.

Time (minutes)	Temperature (°C)	
	Dry tube	**Wet tube**

3 Fill both test-tubes to the top with hot water from the same source. This ensures that the water in both tubes is at the same temperature.

4 Measure the temperature of the water in each of the tubes and record these in the results table at time 0 minutes.

5 Measure the temperature of the water in each tube at intervals of one minute for the next 15 minutes and continue to record these in the results table.

MAKE SURE THAT YOU GET WHAT EVER ASSISTANCE YOU MAY NEED TO MEASURE THE TEMPERATURES OF THE WATER IN THE TUBES AND/OR TO RECORD THE RESULTS IN THE TABLE.

6 When you have finished your experiment, plot your results on a graph with the axes shown below. Put both lines on the same graph.

MAKE SURE THAT YOU GET WHATEVER ASSISTANCE YOU MAY NEED TO PLOT YOUR GRAPH.

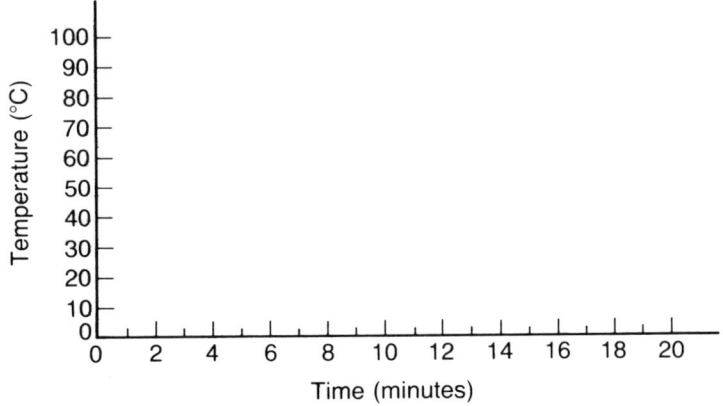

7 From your table, what was the temperature of the water in the dry tube at 8 minutes?

8 From your graph, what was the temperature of the water after 4½ minutes?

9 Describe what pattern you see in the results for the wet tube.

10 How does the pattern of the results for the dry tube compare with the results for the wet tube?

11 What conclusion do you draw from these results?

12 Explain why there is a difference between the water in the tubes in how quickly the water cools down.

13 What would the temperature be in the dry tube after 17 minutes?

Teacher's notes

Assessment objective

Designing an experiment to test a hypothesis
(NEA skill 32)

Previous experience

The nature of the drug to be tested in this exercise
means that the exercise is likely to be used after
consideration has been given to the circulatory system
and its function. Pupils must understand what blood
pressure is and an understanding of why high blood
pressure is dangerous would be necessary.

Management strategy

This is an exercise in which pupils write answers to
written questions.

The pupils' sheet is self explanatory and the activity
needs little setting up. Naturally, assistance should be
provided to pupils having difficulty so that they can be
successful in later questions and complete the exercise.
This assistance would be most easily provided by
having "help" cards or sheets available giving answers
to each of the questions on the pupil sheet. The pupils
requiring "help" cards would be recorded since the
need for help indicates inability to achieve competence
in the skill.

The exercise could be used across the whole
attainment range, although since it does not give pupils
the opportunity to suggest hypotheses it could be used
with groups of average and below average attaining
pupils to assess most of the skills suggested in the
performance criteria in Chapter 2 for this objective.

Time required   ½–1 hour

Resources required

No resources are necessary for this exercise.

Marking the assessment

The marking points below represent suggested answers
to the questions which should be given if pupils are to
be deemed competent in each of the skills. If a pupil
achieves competence in a skill, a "tick" is placed
against that pupil's name for that skill in the master grid
for this objective.

Questions 1, 2, 3 and 6 test the ability to plan a
suitable procedure. Questions 4 and 5 test the ability to
suggest variables and take steps to control them.
Questions 7, 8 and 9 test the ability to suggest suitable
measurements, their frequency and how the pupil will
know whether or not the hypothesis has been proved
correct or false.

Suggested answers:

Q1 People with high blood pressure.

Q2 People also with high blood pressure.

Q3 Any large number (eg ⩾ 100 per group).

Q4 Weight, age, (any others suggested which are
reasonable)

Q5 Choose people of the same weight and same age.

Q6 Suggestion of reasonable method to administer
drug
Statement that all would receive the same amount
of drug
Control group given suitable placebo.

Q7 Blood pressure.

Q8 Suitably frequent time intervals, eg every week.

Q9 If drug has worked, blood pressure in test group
will be reduced but in control group it will not have
reduced.

Exercise 13   Testing a New Drug

Introduction

When a drug manufacturer has designed a new drug to treat some form of disease it is important that it is properly tested to make sure that it will do what it is supposed to and that it is safe for people to use. Usually this testing starts with animals and, if all goes well, a trial is set up using humans. In this exercise you have to try to design a trial for testing a new drug by answering the questions written below.

The drug

Many people suffer from high blood pressure for many different reasons. High blood pressure is dangerous so drugs are used to treat such people to reduce their blood pressure to the normal level.

A new drug called PRESDON has been developed by a drug company to lower blood pressure in people whose blood pressure is too high.
Presdon has been through trials with animals and the drug company is ready to test it in humans.

YOUR TASK IS TO DESIGN A TRIAL TO TEST THE DRUG TO FIND OUT IF IT DOES REDUCE BLOOD PRESSURE IN PEOPLE WITH HIGH BLOOD PRESSURE.

The hypothesis

The hypothesis you are therefore going to test is:

THE DRUG PRESDON CAN REDUCE BLOOD PRESSURE IN PEOPLE WHO ARE SUFFERING FROM HIGH BLOOD PRESSURE.

The trial – your task

Write down your answers to these questions.

1 What sort of people would you use to test the drug on?

2 What sort of people would you use as a control group?

3 How many people in each group would you choose?

4 What factors would you have to keep constant in your test and control groups of people?

5 Suggest how you would keep each of these factors constant.

6 Now that you have your two groups of people and have controlled all of the factors, describe what you would do to each group to test the drug fairly.

7 What measurements would you make?

8 How often would you make the measurements?

9 How will you know if the drug has or has not worked?

Teacher's notes

Exercise 14

Assessment objective

Ability to interpret results
(NEA skills 20, 21, 22, 25 and 26)

Previous experience

It is necessary for pupils to be aware of the need of plants for mineral ions for growth. It may be considered desirable also for them to be aware that if the concentration of ions in the soil is too high, then water will be lost from the plant by osmosis, which explains why in very high fertiliser concentrations the plants grow very little if at all. It is not intended, however, that this explanation is necessary for pupils to be able to satisfy the performance criteria for the highest level of attainment.

Management strategy

Demonstrations

Pupils are presented with the 5 trays set up as suggested below, and are given the sheets. It is possible that there will be more variation between plants in the same tray than between trays. It is advisable when dealing with seeds that account is taken of this inherent variability, and several sets of trays should be set up, the trays showing the 'best' results being used for the demonstration.

While the pupils are copying down the results table, the teacher can measure the heights of the five plants in each tray and write these on an overhead projector or blackboard for pupils to write into their tables. Any help required in doing this should be given.

Pupils can then plot their graph under test conditions and go through the tasks in the 'Interpreting the Results' section of the sheet. This is the part of the sheet which is used to assess their ability in this objective and should also be completed under test conditions. Again, any help needed in doing this section should be given to ensure that pupils are able to complete it if the exercise is used across the whole attainment range. Nevertheless, the section tests all of the skills suggested for this objective in the performance criteria in Chapter 2, and therefore some of the skills are likely to be too difficult for some pupils. Teachers may therefore choose to use the exercise in this format only for average and higher attaining pupils and present it to lower attainers with some of the questions removed. Alternatively, 'help' sheets or cards could be made available with the answers to the questions on them.

Time required   ½–1 hour

Resources required

The suggested application rates of fertiliser are, of course, open to modification as desired. Teachers may prefer to present pupils with applications measured in grams, rather than spatulas.

5 small seed trays containing potting compost treated as follows:

Tray 1: No fertiliser
Tray 2: 1 spatula of fertiliser
Tray 3: 2 spatulas of fertiliser
Tray 4: 3 spatulas of fertiliser
Tray 5: 4 spatulas of fertiliser

Any commercially available solid compound fertiliser will suffice, eg 'Phostrogen'. Alternatively a liquid fertiliser could be used if desired.

Plants should watered as necessary and left to grow to a suitable height for measurement, eg 2–3 weeks.

Marking the assessment

By marking pupils' answers to the questions testing their ability to carry out the skills suggested for this objective, a grade can be awarded for the objective "ability to interpret results". Each question tests a different skill; therefore a pupil would be awarded a 'tick' against any skills which he/she has achieved in the master grid for this objective.

Some of the answers below are necessarily vague since they depend on the actual experimental situation set up and the graphs produced by the pupils. It is important that pupils are not penalised for poor graphs, although these should be rare since they can be given help in drawing them.

Suggested responses:

Q1 Method correct. Value correct +/− 5 mm.
(Ability to perform simple calculations including averages)

Q2 Method correct. Value correct +/− 5%.
(Ability to perform simple calculations including percentages)

Q3 1 spatula.
(Ability to read corresponding values from tables)

Q4 Value correctly read from pupil's graph +/− one division of graph paper.
(Ability to extrapolate information from a graph)

Q5 Value correctly read from pupil's graph +/− one division of graph paper.
(Ability to interpolate data from a graph)

Q6 Correct description of pattern.
(Ability to identify patterns in results)

Q7 Increasing fertiliser application increases growth of plants.
(Ability to draw conclusions from results)

Q8 Explanation of pattern in terms of increased number of mineral ions (or nutrients) in soil needed for growth of plant leading to increased growth in most plants.
(Ability to explain conclusions in terms of underlying theory)

Exercise 14   How does the Amount of Fertiliser Affect the Yield of a Crop?

Introduction

A very important question which farmers need to have answered is "How will the amount of fertiliser that I put on my crop affect its yield?" You will probably realise that this is an important question since fertiliser is expensive and the farmer does not want to use any more of it than is absolutely necessary to get the best yield. Also, putting too much fertiliser on the crop can harm the environment because it can drain from fields into streams, rivers or lakes.

In this exercise you are going to be shown the results of an experiment which has been carried out to try to answer this question.

The experiment

The diagram below shows how the experiment was set up. Five trays were set up with the same number of seeds in each, ie 5. Each tray had a different amount of fertiliser in it. Your teacher will show you the trays that were set up.

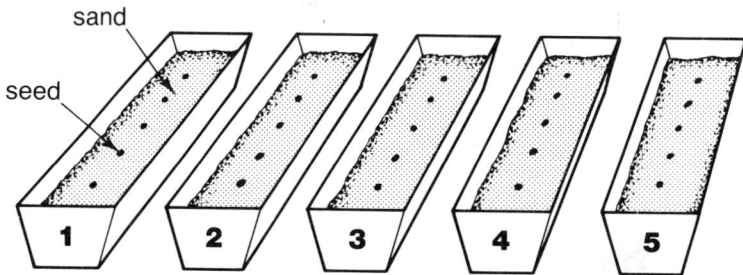

Your task

1 Copy down the results table below:

Tray	Amount of fertilizer In tray (spatulas)	Heights of 5 Plants (mm)					Average Height (mm)
1	0						
2	1						
3	2						
4	3						
5	4						

2 Your teacher will measure the heights of the plants and give you this information. Put this information into the table.

3 To see clearly what effect the fertiliser has had on each crop you should plot a line graph. Draw the axes shown below on graph paper and fill in the graph.

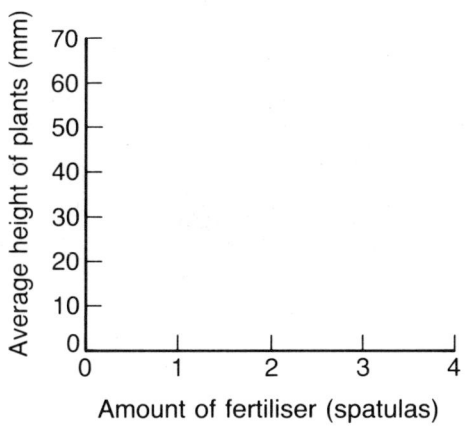

MAKE SURE THAT YOU GET WHATEVER HELP YOU MAY NEED FROM YOUR TEACHER TO DO THIS.

Interpreting the results

Write down the answers to these questions.

1 Calculate the average heights of the plants in each of the five trays and put this information into the table. Show your working underneath the table.

2 Look at the information for the plants in the tray with the least amount of fertiliser in. What percentage of plants in this tray are bigger than the average height? Show your working underneath the table.

3 From the table, what was the smallest amount of fertiliser that was added to any tray?

4 If a sixth tray had been set up, it would have had 5 spatulas of fertiliser on it. Use the graph to predict what the average height of the plants would have been in this tray.

5 What would have been the average height of plants in a tray with 2½ spatulas of fertiliser on it?

6 Study the graph. Describe any pattern you see in the results.

7 What conclusion do you draw from these results?

8 Explain what the fertiliser has done to the soil to have this effect.

Teacher's notes

Exercise 15

Assessment objective

Ability to handle apparatus and materials
(NEA skill 11)

Previous experience

This is an exercise which leaves a record.

Pupils need an understanding of diffusion and an understanding of the function of the kidneys in a mammal in cleansing the blood of unwanted chemicals. This will ensure that they appreciate the importance of the kidney machine for patients who rely on them. It will also be an advantage for them to be familiar with the simple tests for the substances which are to be tested for.

Management strategy

It is intended for the model to demonstrate the principle of dialysis used in kidney machines, rather than to give an accurate reflection of the complex chemicals and barrier systems inside kidney machines. Pupils are therefore given a mixture which is designed to include substances they can easily test for. In addition the substances have been chosen so that some will diffuse out and some will not, thus demonstrating the principles of dialysis.

Pupils can set up the tubing, and while it is being left for diffusion to take place, they should be familiarising themselves with the chemical tests that are to be carried out. It is intended that each pupil sets up a dialysis model but pupils could work in pairs to ensure that they complete all of the tests in the time available. It will be necessary for pupils to work as fast as possible when testing the dialysate; therefore it is important that there are enough reagents for carrying out the tests so that at least a group of pupils working on the same bench have access to a set. One set between two pupils would be even better. If time is available during the 20 minutes' dialysis, pupils could write an account of the exercise.

The assessment is based on pupils' ability to set up the dialysis tubing successfully. It assumes that one of the skills included in the performance criteria for this objective is the assembly of a selectively permeable membrane for use in dialysis or an osmometer. The amount of writing on the sheet may preclude the use of the activity for the very lowest attainers, but this could be overcome with verbal introduction and instructions; whether or not it is felt that the activity itself should be used across the whole attainment range depends on whether this piece of apparatus is regarded as one which requires a high, medium or low level of manipulative skill to set up. It is suggested that it requires at least a medium level of skill and would therefore be used with the full range of pupils and those with higher level of manipulative skill.

Time required   Approx. 1 hour

Resources required

Per pupil:

5–10 cm length of Visking tubing
Distilled water
Cotton thread
Test-tubes
Test-tube holder
Test-tube rack
Bunsen burner and bench mat
Measuring cylinder/5 cm^3 pipette or syringe.
250 cm^3 beaker

Available to each pupil:

"Mixture" of substances in solution containing:
- egg albumen
- starch
- glucose
- sodium chloride
- sodium hydrogencarbonate
Benedict's or Fehling's solutions
Biuret solutions
Iodine in potassium iodide solution
Dilute hydrochloric acid
Dilute nitric acid
Dilute silver nitrate solution

Marking the assessment

The marking points below show what is required in order for students to be awarded competence in the skill of handling apparatus in this exercise. If a student achieves all of the points listed below then he/she will be awarded a "tick" for this skill in accordance with the performance criteria for "handling apparatus and material".

- watertight knot tied in one end of dialysis tubing;
- watertight seal in top of tubing tied with cotton thread;
- tubing filled so that it is firm;
- tubing completely submerged with water in beaker.

Exercise 15   How do Kidney Machines Work?

Introduction

The picture below shows a kidney machine.

People whose kidneys which don't work properly often have to be connected up to this machine overnight. During this time the kidney machine removes all of the dangerous waste chemicals in the blood which are normally removed by the kidneys.

How does it work? In this exercise you are going to make a simple model of a kidney machine to show you how it works. The process is called DIALYSIS.

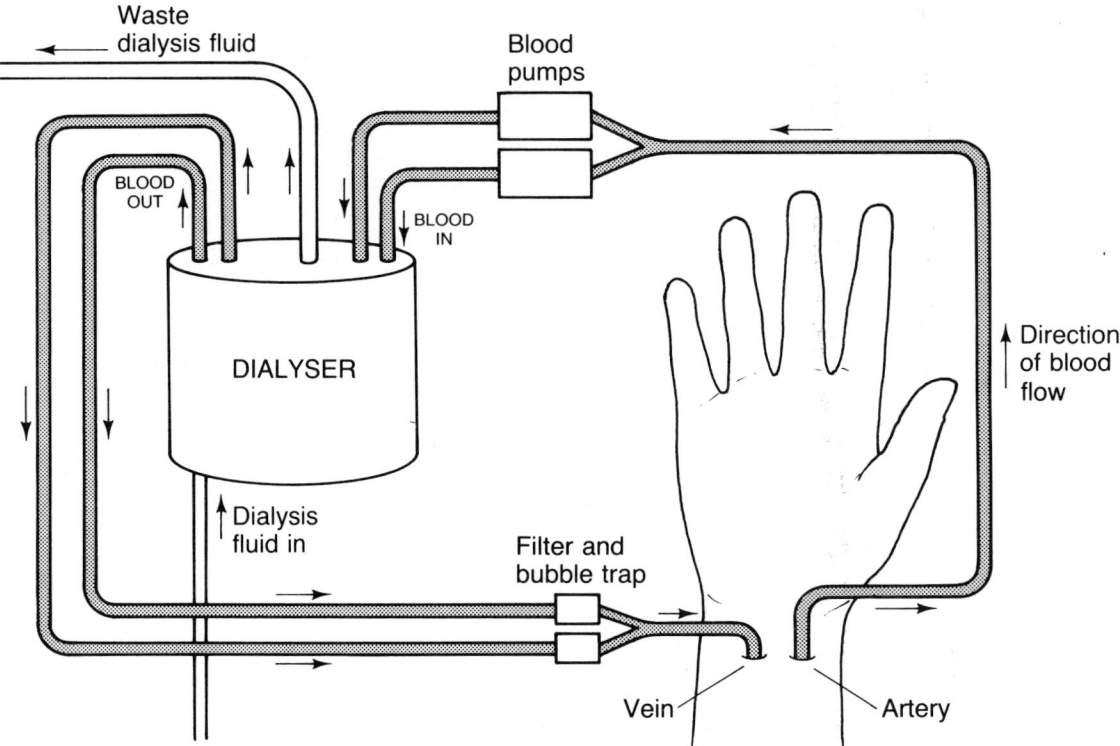

The model

The diagram shows you the set up for the model kidney machine. The tubing is called SELECTIVELY-PERMEABLE tubing, which means that it will let some substances pass through it but not all. This is because it has tiny holes in it which are big enough for some particles to pass through but too small for some others. It is the large particles which must be kept in the blood and most of the small ones which must be removed. If you put a mixture of substances inside it, only those which have very small particles will get out of the tubing. These diffuse out into the water outside the tubing, leaving the substances with big particles inside the tubing.

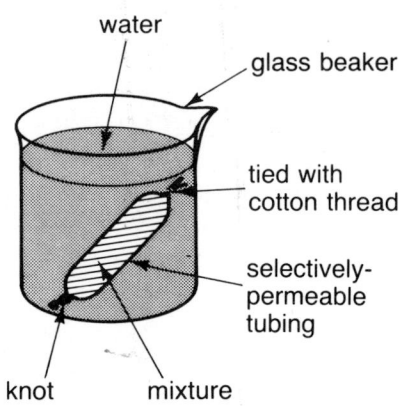

Setting up the model

These instructions tell you how to set up the model kidney machine. The tubing will contain a mixture of substances with different sized particles. The mixture is not intended to be exactly the same as blood. It is intended to show you how dialysis works in the kidney machine. For this reason substances have been chosen which you are familiar with and which you can test for. The substances in the mixture are:

– protein
– starch
– glucose
– sodium chloride
– sodium hydrogencarbonate

1 Get a piece of tubing and tie a knot in one end.

2 Fill up the tubing with the solution which contains the mixture of substances.

3 Tie up the top of the tubing with cotton thread so that the tubing is quite firm.

3 Wash the outside of the tubing with distilled water to clean off any solution which may be on the outside.

4 Place the tubing into the beaker of distilled water and leave it there for 20 minutes.

5 While you are waiting for the dialysis to take place, get together test-tubes in a test-tube rack and the chemicals you will need to do the tests described below. Read through the tests and make sure that you know what you are going to do at the end of the 20 minutes.

Testing the model

The diagrams below give you instructions for testing the water around the tubing to see which substances have diffused out.

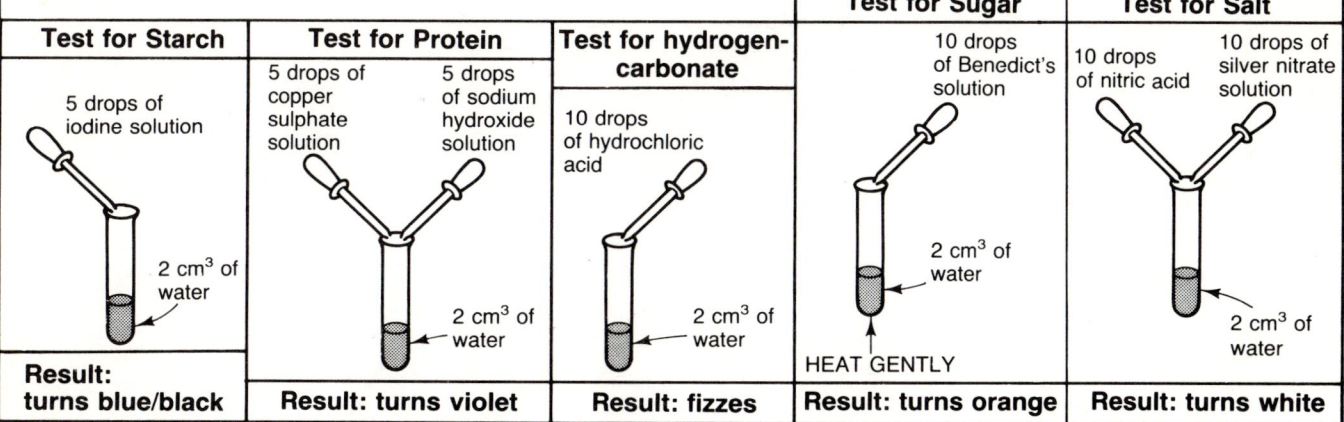

Carry out the tests on 2 cm³ of the water in the beaker.

Keep the tubing for your teacher to see.

Teacher's notes

Assessment objective

Ability to observe: fine detail
        differences in fine detail
(NEA skills 8 and 10)

Previous experience

No previous experience is essential, although it would be advantageous for pupils to have been made familiar with the layers of tissue which can be found within a leaf so that they know which layers they should be examining in this exercise.

Management strategy

This is an exercise which can be assessed during routine coursework.

Pupils will examine the transverse section of a leaf, describe the features they observe in the palisade mesophyll and describe differences between the palisade and spongy mesophyll tissues. It is intended that this activity should be used with pupils of average and above average attainment since it has been suggested that the skills "describing fine detail" and "describing differences in fine detail" are high-level observation skills.

This is intended to be a short activity so that a proportion of the pupils can be assessed during the lesson while the remainder are engaged in other activities.

Time required   Up to ½ hour

Resources required

Per pupil to be assessed:

Microscope with high power objective lens
Bench lamp (if necessary)
Slide of leaf, TS

Marking the assessment

It is necessary for the teacher to draw up a set of marking points for the exercise since these are dependent upon the slides which the pupils have available to observe. The guide below outlines the procedure which should be followed, with suggestions of observations for which the pupils would be given credit.

The ability to observe fine detail

(a) Draw up a list of fine details which pupils are expected to observe in the palisade mesophyll, eg:
  - cells are long and thin
  - have many green structures/chloroplasts inside
  - have a large space/vacuole occupying most of the volume of the cell
  - have a nucleus
  - have cytoplasm which looks very 'grainy'

(b) Decide how many of these must be observed for a pupil to show competence in this skill.

(c) Mark pupils' work and award a 'tick' in the master grid for this objective for the skill of observing fine detail for any pupils who demonstrate competence by observing the required number of features.

The ability to observe differences in fine detail

(a) Draw up the list of fine differences which pupils should be expected to observe between the palisade and spongy mesophyll layers, eg:
  - palisade mesophyll cells are long and thin; spongy mesophyll cells are much rounder
  - palisade mesophyll cells are packed closely together side by side; spongy mesophyll cells are packed much more loosely
  - palisade mesophyll cells have many more chloroplasts than spongy mesophyll cells
  - spongy mesophyll cells have large gaps between them; palisade cells have much smaller gaps.

(b) Decide which of these must be observed to be awarded competence in this skill.

(c) Mark pupils' work and award a 'tick' in the master grid for this skill for any pupils who demonstrate competence.

Exercise 16   How is a Leaf Built to Do its Job?

Introduction

The diagram below shows a transverse section through a leaf.

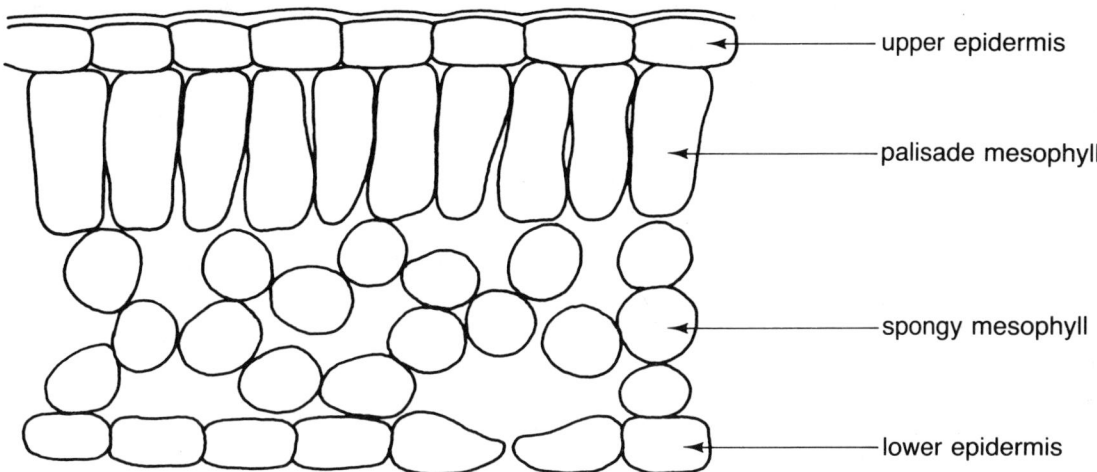

upper epidermis

palisade mesophyll

spongy mesophyll

lower epidermis

The leaf is the organ of a flowering plant which is designed for making food by photosynthesis. The leaf is made of several layers of tissue and these have been labelled in the diagram above.

It makes sense that if a leaf is designed to carry out photosynthesis then it must be built so that it can do this job as well as possible. In this exercise you are going to investigate how well a leaf is built to do its job.

Instructions

1 Set up your microscope so that you can examine the transverse section of a leaf that you have been given.

ASK YOUR TEACHER TO HELP YOU DO THIS IF NECESSARY.

2 Identify the PALISADE MESOPHYLL in your slide using the diagram above to help you.

YOU CAN ASK YOUR TEACHER FOR ASSISTANCE IF YOU WISH.

3 Examine this layer of tissue very closely. Describe the palisade mesophyll in as much detail as you can.

4 Now identify the SPONGY MESOPHYLL layer using the diagram above to help you.

YOU CAN ASK YOUR TEACHER FOR ASSISTANCE IF YOU WISH.

5 Now describe as carefully as you can any differences you can observe between the palisade mesophyll and the spongy mesophyll.

Teacher's notes

Assessment objective

Ability to handle materials and apparatus
(NEA skill 13)

Previous experience

No previous experience is essential, therefore this
exercise might be used as an introductory activity in a
topic investigating reproduction in flowering plants.
However, a familiarity with flower structure may be an
advantage, particularly to lower attaining pupils, since
the successful completion of the assessment depends
upon pupils having a very clear understanding of what
must be dissected out of the flowers.

Management strategy

An additional assessor is suggested for this exercise.

The exercise is intended to assess pupils' ability to
use dissection instruments and therefore assumes that
this skill has been stated as one of the skills listed in the
performance criteria for this objective.

To make most efficient use of the additional assessor, it
would be expected that the teacher would attempt to
assess all pupils on this objective. To that end, the task
has been kept simple, although different pupils will
take different lengths of time to complete it.

Pupils would be given the flower and then follow the
instructions on the sheet. Clearly it is important that
flowers are provided which have very distinct floral
parts to minimise confusion over identification of the
parts. It is most important that the ability of pupils to
dissect out the parts of the flower is not prejudiced by
their understanding of which parts are which. They
must therefore be given any assistance that is required
in identifying the parts of the flower.

It is intended that the task is used across the
attainment range.

Time required   20 minutes

Resources required

Per pupil to be assessed:

Simple flower, eg Tulip
Scalpel
Forceps
Mounted needle
Blunt seeker
Bench mat
Hand lens (optional)

Marking the assessment

The marking points suggested below are those which a
pupil must satisfy in order to attain competence in the
skill of using dissection instruments within the performance criteria for the objective of "handling apparatus
and materials". If pupils achieve success in this skill
they would then be awarded a "tick" for this skill in the
master grid for this objective.

How many of these criteria must be satisfied in
order to gain competence is a matter of prefessional
judgement. It is suggested that pupils must satisfy them
all in order to achieve competence.

– sepals removed cleanly and without damage
– sepals displayed neatly
– petals removed cleanly and without damage
– petals displayed neatly
– stamens removed cleanly and without damage
– stamens displayed neatly
– carpel removed cleanly and without damage
– carpel displayed neatly

Exercise 17   Flower Structure

Introduction

Flowers are the organs which plants use to produce their seeds so that they can breed (or 'reproduce'). Different parts of the flower do different jobs to make the seeds. To understand how plants make their seeds, we must first understand what the parts of the flower are. This exercise is designed for you to do this.

Below are two diagrams of the same flower. The one on the left is the whole flower and the one on the right is an exploded view of the same flower. This shows what the parts of the flower are called.

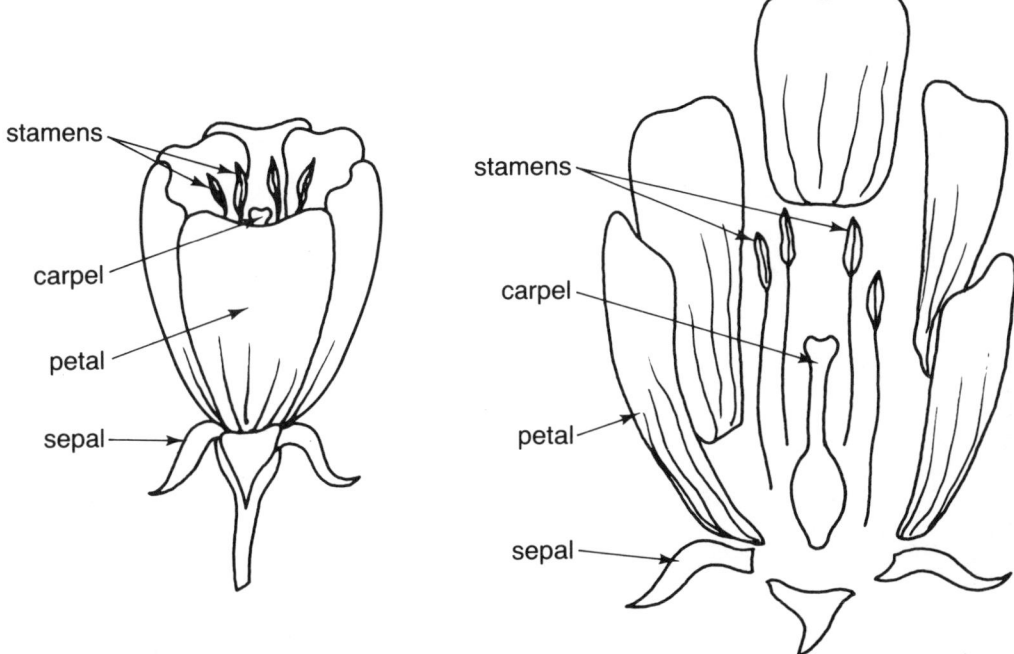

You will have a flower to investigate. It may not be exactly the same type of flower as the one shown above, but this does not matter because all flowers have basically the same parts.

Instructions

1 When you have got your flower, look at it carefully and make sure you can find the following parts, starting from the outside of the flower.

   ASK YOUR TEACHER FOR ANY HELP YOU NEED TO DO THIS.

   Sepals: These are the outer layer of small, leafy, usually shrivelled, parts.

   Petals: The large, coloured parts which are the next layer in.

   Stamens: Several long, thin parts growing out of the base of the flower. At the end of each stamen is a swelling.

   Carpel: The central part. There is usually only one of these.

2 When you are sure that you know which parts are which, you are ready to remove them from the flower.

3 Use the dissecting instruments to carefully remove the four types of parts.

4 Write the words SEPALS, PETALS, STAMENS and CARPEL on a sheet of paper and place each part next to its name.

ASK YOUR TEACHER TO CHECK THIS WHEN YOU HAVE FINISHED.

Teacher's notes

Exercise 18

Assessment objective

Ability to measure: with a metric ruler
with a balance
(NEA skills 1 and 3)

Previous experience

It is intended that this activity might be used to introduce a topic on variation and inheritance; therefore it is not essential that pupils have any previous experience. Nevertheless it might be thought advantageous for pupils to have covered work on sexual reproduction in plants and mammals so that they can understand why there should be variation.

Management strategy

The exercise is a circus of stations at which pupils make measurements and record their results on large histograms, ideally fixed to the wall. The whole class can undertake the activity and all record their results by placing a token (a small paper square with an appropriate symbol on it will suffice) on the histogram in the column which is appropriate to their measurement or observation. If pupils place their tokens one above another, then a frequency distribution histogram will be constructed during the exercise for each of the features being investigated.

The teaching objective is to introduce variation. A teacher may wish to go further and make the point that, since all of the measurements made are of features which show a range of values, this is called continuous variation. It will be necessary to make it clear to pupils that although the histograms have discrete categories, these are only sub–divisions of a continuous scale. Confusion may otherwise arise through pupils failing to understand how the categories of the histograms show examples of continuous (cf. discontinuous) variation.

Assessment of the ability to measure is carried out by the teacher observing pupils carrying out the measurements. It is unlikely that all pupils will be seen since this is a relatively short activity. The teacher would attempt to see as many pupils measuring with these instruments as possible during the course of the exercise. To this end, several of the exercises require pupils to measure with a ruler, thus providing several opportunities to assess pupils.

It is an exercise which is intended for use across the whole attainment range. However, the level of attainment required to carry out the skill of measuring with a balance depends on the balance in common use in the department. If it is a digital balance then this might be listed as one of the skills which pupils showing the lowest level of attainment would be required to be able to use. A top-pan balance would require a higher level of attainment to successfully measure with it. Therefore in such circumstances teachers may choose not to include the weighing activity with groups of lower attaining pupils.

Time required   Approx. ½ hour

Resources required

Human height station

- two metre rulers fixed to the wall, one above the other,
- a large histogram to record heights of pupils as a frequency distribution diagram (see below)
- tokens for recording height

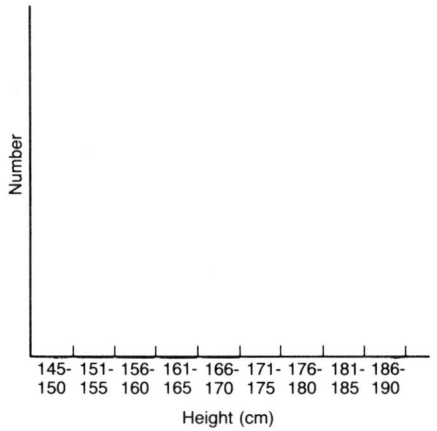

**Illustration of type of histogram needed
(example given is for human height)**

Human finger length station

- metric ruler
- large histogram as described above for recording finger length
- tokens for recording finger length

Plant leaf length station

- selection of leaves from the same plant or species
- large histogram as above for recording plant leaf length
- tokens for recording length
- box to put leaves which have been measured to prevent re-measuring the same leaves

Plant seed weight station

- selection of seeds of the same type (eg broad beans)
- balance
- large histogram to record seed weight
- tokens for recording weight of seeds
- box to put seeds which have been weighed to prevent reweighing

Plant height

- tray of plants of same age (e.g. cereals or beans)
- metric ruler
- large histogram for recording height
- tokens for recording height

Marking the assessment

It is suggested that pupils must satisfy the marking points below in order to achieve competence in the use of the metric ruler and balance. Any pupils achieving competence would be awarded a 'tick' for these two skills in the master grid of skills for the objective of measuring.

Metric ruler
Measured to an accuracy of within one mm.

Balance
Measured to an accuracy of one division on the scale.

Exercise 18   Variation in Humans and Plants

Introduction

Look at your neighbour. How many features do the two of you have in common? How many features are not the same?

Plants and animals of all species are the same as this: they all have some features which are the same for all of them and some which are different for different members of the same species. Differences between members of the same species are called VARIATION.

In this exercise you are going to investigate variation in plants and animals of the same species. Your teacher will show you which plants and which parts of plants you will be looking at. The animals you will investigate are the people in your class.

Go around the room and at each of the stations follow the instructions for that station which are written below.

WHILE YOU ARE DOING THE TASKS AT EACH STATION YOUR TEACHER WILL ASSESS YOUR ABILITY TO MEASURE.

Human height

1 At this station, measure your height with the ruler stuck to the wall.
2 Place one of the tokens on the bar chart which has been stuck on the wall in the correct column for your height.

Human finger length

1 Measure the length of the middle finger of your right hand with the ruler at the station.
2 Place one of the tokens on the bar chart which has been stuck on the wall in the column for your finger length.

Plant leaf length

1 Measure the length of one of the leaves on the bench at this station and then put the leaf in the box.

2 Place one of the tokens on the bar chart which has been stuck on the wall in the column for the length of the leaf you have measured.

Plant seed weight

1 Measure the weight of one of the seeds on the bench with the balance and then put the seed into the box.

2 Place one of the tokens on the bar chart which has been stuck on the wall in the column for the weight of the seed you measured.

Plant height

1 Measure the height of one plant from each tray at this station.

2 For each type of plant, put one of the tokens onto the appropriate bar chart on the wall in the correct coloumn for the height of that plant.

Teacher's notes *Exercise 19*

Assessment objective

Ability to observe: gross features
                     differences in gross features
(NEA skills 7 and 9)

Previous experience

An understanding of the function of the heart is important. It is desirable that pupils have not previously seen any drawings or diagrams of heart structure since this is likely to give them preconceived ideas about what they should be observing.

Management strategy

Demonstration

    This exercise is intended as an activity which is carried out during part of a lesson on the heart and is therefore a short activity, although 1 hour has been suggested as the time required because it is likely to take that long to assess at least half the class. The class would be engaged in other activities, for example watching a video or film showing how the heart works and then making a set of annotated diagrams to illustrate how it works. Individual pupils would then be called to look at the heart itself and carry out the exercise. One heart, of course, can provide two sections so it is possible for at least two pupils to be carrying out the activity at once. Clearly the more hearts that are available, the greater the number of pupils that can be assessed: one heart per pupil would allow this exercise to be used with the whole class at once. After all the pupils have done the exercise, a discussion can take place linking the structure of the heart to its function.

    Differentiation in this objective is largely achieved by cueing. Instruction 1 is necessarily very open-ended since the highest achievers should be given the opportunity to be successful in observing without cueing. Lower achieving pupils, however, may need to be cued to be successful. The teacher is on hand to provide this assistance by giving examples of major features which can be observed, eg 4 large holes/pockets/chambers in the heart. With this assistance available, the exercise is intended for use across the whole attainment range.

Time required   Approx. 1 hour

Resources required

One heart cut longitudinally (the larger the better).

Marking the assessment

Exactly what a teacher can expect pupils to observe depends, of course, on what is clearly visible in the actual heart being examined. In preparation the teacher must therefore list the marking points for the two skills which pupils can reasonably be expected to observe from the particular heart(s) used in the exercise. Examples of major features which may be observed are:

- heart made of red tissue
- heart made of thick tissue
- white material (fat) around parts of the heart tissue
- four holes/pockets/chambers in heart (possibly only 2 if atria have already been cut off as is often the case)
- long threads inside these chambers
- tubes leading out of chambers

    The assessor must decide how many of the features in the mark list must be observed to demonstrate competence.

    Pupils' lists of observations would then be marked using these mark lists. Any pupils who achieve competence in these skills would then be given a 'tick' for one or both of these skills in the master grid for the objective of observing.

Exercise 19   Heart Structure

Introduction

The mammalian heart is an amazing organ! It can pump blood around the body, without ever stopping, for over 100 years!

If it is to pump blood around the body it must be specially built to do so. In this exercise you will have the opportunity to look closely at how the heart is constructed. You can then think about how this structure is suited to the job it does.

Instructions

Your teacher has a heart which has been cut open. When you have the opportunity to look at the heart you should do the following things:

1 Look at the whole heart. Pay particular attention to main features of the heart. Write down a list of all of the main features that you observe.

2 Your teacher will show you the left and right ventricles. Study them both carefully and write down any differences you observe between them.

Teacher's notes

Assessment objective

Ability to follow instructions
(NEA skills 11 and 12)

Previous knowledge

No previous knowledge is essential although since this exercise is likely to be used within the context of work on factors in the habitat which affect where organisms live, it would be advantageous if pupils already understand that there are several environmental factors which influence where organisms live. This will provide a more meaningful point to the exercise.

Management strategy

This exercise allows for assessment during routine coursework and can take as long or short a time as is required. Pupils may be expected to repeat their experiments to provide, say, three replicates, in which case as many as half a class may be assessed. However, the management strategy suggested assumes that only a small proportion of a class, say 5 pupils, will be assessed since it is clearly necessary to watch pupils as they are carrying out the experiment.

This exercise provides opportunities for pupils to follow a single instruction and a sequence of instructions. It can therefore be used to assess up to the attainment level which is capable of following sequences of instructions. If the teacher is observing the pupils as they follow the instructions then differentiation can be achieved by providing pupils with assistance as required. For example reading the instructions out to allow the pupil to follow them one at a time if the pupil is a low attainer.

Time required   Approx. 1 hour

Resources required

Per pupil being assessed:

Choice chamber
This consists of two plastic petri dishes welded or glued together with a piece cut out of the top pieces where they join so that woodlice can get through the gap. (See pupil sheet).

10 woodlice
water at room temperature
metal gauze (to fit between top and bottom of choice chamber)
silica gel or anhydrous calcium chloride
black paper or cloth (big enough to cover half of choice chamber), or half of choice chamber painted black
stopclock

Marking the assessment

To assess the ability to follow a single instruction the teacher would observe a pupil while he/she is actually carrying out a single instruction from the pupil sheet. If a pupil can work through all of the instructions for either of the experiments then the pupil has achieved competence in following a sequence of instructions.

Faster pupils could be given the task of carrying out the experiment that they design in answer to question 3 in the section entitled 'Interpreting the Results' which gives further time for the assessment of slower pupils if desired.

Exercise 20   What Conditions do Woodlice Prefer to Live in?

Introduction

Woodlice are usually found in dark, damp places. It seems likely, therefore, that, given the choice, they prefer to live in dark, damp places. In this exercise you are going to test whether or not this is true.

The experiment

To do the experiment you need a piece of equipment called a CHOICE CHAMBER. The diagram below shows what a choice chamber looks like. It is a chamber divided into two equal sized portions with a gap in the partition joining the two portions together. It is possible to set up the choice chamber so that the conditions in the two sides are different. If you put woodlice into the chamber they will go to the side with the conditions they prefer.

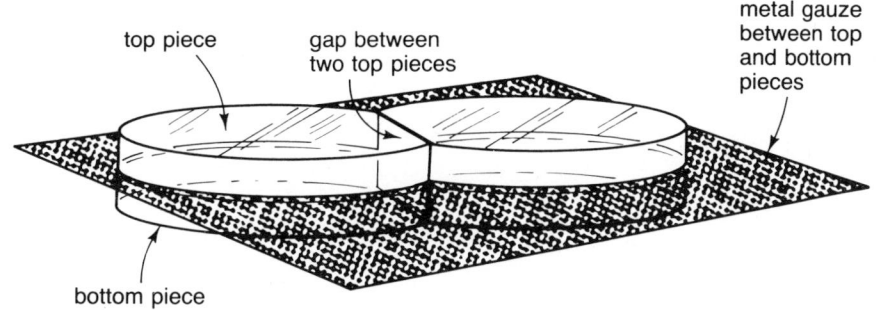

Follow the instructions below to find out whether it is true that woodlice like to live in dark and in damp conditions.

Instructions

Test for darkness

Before you start the experiment you need to draw a table for your results. Include a column for time (in seconds) and a column for number of woodlice.

1 Place the metal gauze over the bottom piece of the choice chamber.

2 Place 5 woodlice on top of the gauze in each side of the choice chamber.

3 Place the top of the chamber over the gauze.

4 Make one the sides of the chamber dark by covering it with black paper or cloth.

5 Start the stopclock.

6 Record how many woodlice are in each of the sides at 30 second intervals for 5 minutes.

Test for dampness

1 Remove the top of the choice chamber, the woodlice and the gauze.

2 Cover the bottom of one side of the chamber with water which has been kept at room temp.

3 Cover the bottom of the other side of the chamber with a thin layer of silica gel (this absorbs moisture out of the air).

4 Place the gauze over the bottom piece of the chamber.

5 Place 5 woodlice on the gauze in each side of the choice chamber.

6 Place the top of the chamber over the gauze.

7 Start the stopclock.

8 Record how many woodlice are in each side of the chamber every 30 seconds.

Interpreting the results

1 Study your results. Does it appear that the woodlice prefer to be in the light or in the dark?

2 Does it appear that the woodlice prefer to be in the damp or dry?

3 How could you test to find out if they prefer being in damp **and** dark conditions compared to just damp or just dark conditions?